SANTA ANA PUBLIC LIBRARY
D0387649

MARIE CURIE

A BIOGRAPHY

MARILYN
BAILEY
OGILVIE

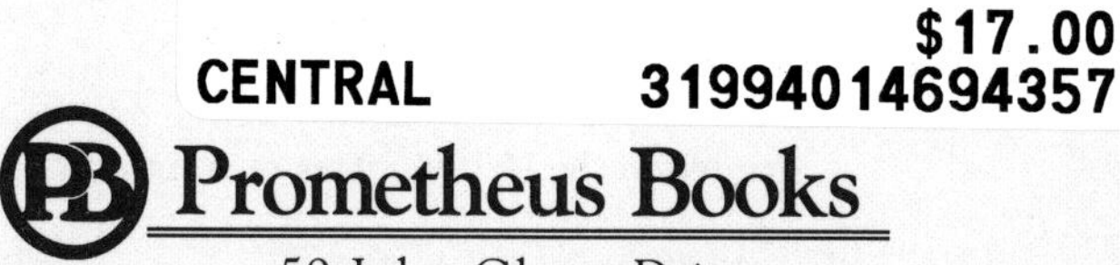

59 John Glenn Drive
Amherst, New York 14228-2119

Published 2011 by Prometheus Books

Inquiries should be addressed to
Prometheus Books
59 John Glenn Drive
Amherst, New York 14228–2119
VOICE: 716–691–0133
FAX: 716–691–0137
WWW.PROMETHEUSBOOKS.COM

15 14 13 12 11   5 4 3 2 1

Library of Congress Cataloging-in-Publication Data

Ogilvie, Marilyn Bailey.
Marie Curie : a biography / by Marilyn Bailey Ogilvie.
p. cm.
Reprint. First published: Westport, Conn. : Greenwood Press, 2004.
Includes bibliographical references and index.
ISBN 978–61614–216–2 (pbk.)
1. Curie, Marie, 1867–1934. 2. Chemists—France—Biography. 3. Women chemists—France—Biography. 4. Physicists—France—Biography. 5. Women physicists—France—Biography. I. Title.

QD22.C8035 2010
540.92—dc22

2010022920

Printed in the United States of America

CONTENTS

INTRODUCTION

When asked to name an important woman scientist, most people would only hesitate a short time before answering, "Marie Curie." The reasons seem obvious. Marie Curie made one of the most important theoretical breakthroughs of the twentieth century when she postulated that radiation was an atomic rather than a chemical property. She was the first person to use the term radioactivity. Her studies motivated a long search that culminated in the isolation of two new elements, polonium and radium. Two aspects of Marie Curie's scientific genius emerge: creativity and perseverance. Although the imaginative discovery of the atomic nature of radiation is perhaps her most significant contribution, without another characteristic, perseverance, she would have been unable to substantiate her hypothesis. Her scientific work netted her two Nobel Prizes, one in physics and the second in chemistry.

When looking at the life of this remarkable scientist, it is easy to picture a stern, one-dimensional woman so totally committed to her science that she was incapable of complex emotions. A deeper examination reveals a woman whose childhood was marred by the sickness and death of a mother and sister, and a father who was also scarred by these losses. Her father struggled to support his remaining four children as a teacher under an oppressive regime in a Poland controlled by the tsar of Russia. Marie's reaction was to reject the religious beliefs of her childhood and to become involved in political movements. Since many obstacles prevented girls from attending universities in Poland, Marie joined an underground, unofficial university.

In order to earn enough money to attend a foreign university, Marie left home to become a governess. She promptly fell in love with the son of her employers. The love affair was a disaster, making Marie wary of any commitment in the future. When she finally met Pierre Curie, she was reluctant to pledge herself to another relationship. Once she decided to entrust her emotions to Pierre, her loyalty was unswerving even after his tragic, premature death. Marie loved their two children, but sometimes emotionally neglected them, as she herself had felt neglected.

After Pierre's death when Marie's friendship with the married physicist Paul Langevin blossomed into love, the entire country was incensed. From a grieving widow, Marie was portrayed as a scheming home wrecker. Duels were fought between her supporters and detractors and scurrilous newspaper editorials bashed her. Exhausted and ill after the controversy, she gradually reentered society. In her later life she spent much of her time working to develop a new research institution dedicated to radioactivity. During World War I, she established a fleet of mobile x-ray units transported in specially fitted cars. After the war, although she had the time to devote to research, money and supplies were absent. In order to supply her laboratory she traveled to the United States twice and undertook a job totally antithetical to her shy public personality. She became an ambassador for science in a role as fund-raiser.

In her later years she made many enemies within the male scientific establishment who disparaged her work and claimed that her early successes were only possible because of Pierre. As her health declined, she went to her laboratory until she finally could do so no longer. She was a beloved mentor to younger scientists at the Radium Institute, which she had pioneered. Plagued by fatigue, cataracts, and acute anemia she courageously went to the laboratory and gave her lectures at the Sorbonne until her final illness. Her beloved radium eventually killed her as it did her husband many years before.

Marie Curie was a very complex person. A fine creative scientist, she was dogged by her personal demons but managed to transform them into successes. In 1935, the year after her death, Albert Einstein published a memoriam to her in which he attributed her discovery of the two new elements both to intuition and to tenacity under the most trying circumstances imaginable. He concluded that of all famous people she was the only one whom fame had not corrupted. It is not surprising that when we think of a famous woman scientist her name comes to the forefront.

A biography is the story of an individual's life. No life is lived in a vacuum, and Marie Curie's life is no exception. By understanding how this outstanding scientist operated within the context of late nineteenth-

and early twentieth-century science and society, we are better able to understand both her life and her science. Any individual is the product of many factors. Each person is influenced by parents and siblings, education, religious background, socioeconomic status, spouse and children, national background, and social and political ideals. Both people and science are central to Marie Curie's story. Her family, friends, and scientific colleagues played an essential role in her life. They both molded her and were molded by her. To begin to understand Marie Curie, we must look at all of these factors. Her scientific achievements became the standard for what a woman could attain in science. In order to truly fathom her unique accomplishments we must look at the achievements of some of her contemporary women scientists. Her science was impacted by the landmark discoveries of other scientists, both male and female, so in order to understand her place in the history of science it is important to consider the achievements of other investigators. This biography consists of eleven chapters, but is basically divided chronologically into three major sections: (1) early life and education in Poland and her work as a governess; (2) the major creative part of her life including her university achievements; marriage to, collaboration with, and death of Pierre; raising children, and major scientific achievements culminating in two Nobel Prizes; and (3) finally the last part of her life where she operated a radiology service during World War I, directed her own radium institute, became a fund-raiser for radium as she had become an international icon, and finally died from exposure to the elements that she discovered.

TIMELINE OF EVENTS IN THE LIFE OF MARIE CURIE

November 7, 1867	Maria Sklodowska born in Warsaw, Poland.
May 9, 1878	Maria Sklodowska's mother died.
June 12, 1883	Graduated from secondary school and obtained a gold medal.
January 1, 1886	Left job with the Zorawskis.
March 1889	Registered as a student at the Sorbonne.
November 5, 1891	Degree in physics from the Sorbonne; first place.
June 1893	Degree in physics from the Sorbonne; first place.
July 1894	Degree in mathematics from the Sorbonne; second place.
July 26, 1895	Married Pierre Curie.
September 12, 1897	Irène (Joliot-Curie) was born.
July 18, 1898	Marie and Pierre announced discovery of polonium.
July 26, 1898	Announced discovery of radium with Pierre and Gustave Bémont.
September 12, 1898	Introduced term radioactivity in a published article.
December 1903	Nobel Prize in Physics, shared with Henri Becquerel and Pierre.
December 6, 1904	Eve Curie was born.
April 19, 1906	Pierre died in an accident.
November 5, 1906	Became first woman professor at the Sorbonne.
January 23, 1911	Failed to be elected to the French Academy of Sciences.
December 1911	Awarded the Nobel Prize in Chemistry.

August 1914	Radium Institute completed.
1914–1919	Operated mobile x-ray unit during World War I.
May–June 1921	Visited the United States to receive gram of radium.
1929	Second trip to the United States to raise money for radium research institute in Poland.
July 4, 1934	Died of aplastic anemia.
April 20, 1995	Reburied in the Panthéon. First woman so honored for her own accomplishments.

Chapter 1

EARLY LIFE AND EDUCATION

The four surviving children of Bronislawa Sklodowska and Wladyslaw Sklodowski gathered around their gravely ill mother on May 8, 1878.[1] Bronislawa's body was ravaged by the tuberculosis she had contracted some time before 1871 when her youngest daughter, Maria (later to be known as Marie Curie), was only four years old. At this time the only known treatments for tuberculosis were rest, medicinal waters, and a healthy climate.[2] Bronislawa spent two years of Maria's childhood away from the family home in Warsaw, Poland, taking the "cure" in a French spa. Even before their mother's sickness and ultimate death, Maria's childhood had been difficult. Some of the family's problems were political. Although Poland had once been one of the largest nations in Europe, a series of wars divided it into three provinces, each controlled by a different country. The Sklodowskis grew up in Warsaw, a city dominated by the Russian tsars. Although Wladyslaw was a loyal Pole, he was a mathematics and physics teacher in a government school under Russian control. This situation made his position in the school very precarious, for if he wanted to keep his job he had to appear to be conforming to Russian requirements. If he allowed the students to speak Polish or used Polish himself, he would be replaced by a Russian, since the Russian police kept a close watch over Polish teachers. And should he be suspected of harboring any revolutionary ideas, he would be arrested and severely punished. The fear of reprisals haunted the Sklodowski children throughout their lives.

Children whose lives are filled with pleasure, joy, and love react very differently throughout their lives from those who are haunted by fear and anxiety. Well-to-do middle-class American children find it difficult to understand what it is like to be scarred by apprehension and hunger. To Iraqi, Rwandan, and Bosnian children or to American inner-city children these conditions are normal, and they live their lives accordingly. Throughout her life, Marie Curie felt a great loyalty to Poland, but the hardships that she endured as a child influenced the kind of adult that she became.

Poland's problems stretched back over many years. With a government that had long been inefficient, chaotic, and corrupt, it was ripe for a takeover by its stronger, more efficient neighbors, Austria, Russia, and Prussia. Eager to increase their own territories and thus their power, Poland was partitioned among these countries in the late eighteenth century. For a short time, the Poles saw powerful Napoleon as a potential benefactor. As he was racing around Europe conquering country after country, Napoleon used Polish Legions in many of his battles. After he defeated Austria and Prussia, Napoleon created the Duchy of Warsaw and helped the Poles raise their own army. At the same time, Russia prepared its own plan for restoring the Polish state under the rule of the Russian tsar, Alexander I. Any hope that the Polish people had of regaining control over their country through Napoleon was squelched after his disastrous defeat by Russia (1811–1812). Napoleon's Duchy of Warsaw was replaced by a Kingdom of Poland, connected to Russia by a union with the tsar of Russia. This tsar also became king of Poland, which had its own constitution, parliament, army, and treasury. The remaining territories were united under Prussian rule.

Constant tension existed between the despotic administration in Russia and the constitutional regime in Poland. Young Poles conspired against the government and planned an uprising. On November 29, 1830, the rebellion exploded. Even though the Polish army fought valiantly, in the end it could not compete with the superior resources of Russia and was forced to surrender in September 1831. After the failure of the revolt, many of the concessions the Poles had previously gained from the Russians were taken away. The constitution was annulled, the army was liquidated, Warsaw University was closed, property was confiscated, and suspected dissidents were deported. Many of the exiled leaders went to France, forming an expatriate community in Paris that Marie Curie would later join.

Another failed uprising occurred in January 1863 and lasted

through 1865. In the earlier rebellion (1830) as in this one, few peasants were involved. Since they had few rights, most peasants were not overly concerned about who ruled them. Most of the rebels were priests, clerks, burghers (merchants), gentry, and intellectuals. The rebellious Poles were so harshly defeated in 1865 that they gave up the idea of liberating Poland through military means for many years and resorted to civil disobedience instead. Both the Poles and the Russians needed the support of the peasants in order to be successful. Russian authorities finally recognized that they must court the peasants if they wished to have Poland running smoothly. In 1864 the Russian tsar issued a decree enfranchising them. This long overdue freedom from their feudal obligations did not have the effect that the Russians expected. Instead, the peasants gradually became members of the National Polish Community—the goal of the rebellious Poles.

It was into this political climate that Marie Curie's parents met and married. When Wladyslaw married Bronislawa Boguska in 1860, she was headmistress of one of the best private school for girls in Warsaw. Bronislawa was the oldest of six children of a family of country squires. Although the family belonged to the landowning nobility, its members had to work for others in order to make a living. The family had enough money, however, to give their daughter a good education in a private school in Warsaw. After graduation, she became a teacher in the same school and eventually became its director. Although she had little money, she was a beautiful, accomplished, well-educated woman who was noted for her musical abilities. Wladyslaw was also part of the minor nobility who, because of the misfortunes of Poland, was poor. He studied science at the University of St. Petersburg in Russia and returned to Warsaw where he taught mathematics and physics. The marriage seemed suitable to all observers, lacking only one thing—money.

The couple moved into apartments adjacent to Bronislawa's classrooms. They lived there for seven years, during which time they had five children: Zofia, known as Zosia (b. 1862); Józef (b. 1863); Bronislawa, named for her mother and known as Bronia (b. 1865); Helena, known as Hela (b. 1866); and Maria, known as Manya (b. 1867). Wladyslaw must have convinced the Russians of his orthodoxy, for he received a promotion, which amounted to a second job that allowed

Bronislawa to give up her position and stay home with her children. He served both as a professor of mathematics and physics, and as school under-inspector. Since they could no longer live in the apartments owned by her school, they moved to quarters provided by Wladyslaw's school. Maria was still a baby when they moved in 1868, but Bronislawa tutored the other children, especially the two older ones, Zofia and Józef.

Because Bronislawa experienced the first symptoms of tuberculosis when her youngest child, Maria, was born, she never held the little girl close or kissed her. Although she may have felt that physically distancing herself from the child would protect her from tuberculosis, it had another less desirable effect on the small girl. Even though her mother smiled and gave her affectionate looks, these signs represented mixed messages to Maria. It was difficult for the youngster to understand the lack of physical contact from a mother who professed a great love for her. Maria worshiped her mother, but when she pushed her clinging hands aside and suggested that she go outside to play in the garden, she felt rejected. These childhood experiences may explain why, as she grew older, Maria found it very difficult to be physically close to people.

Maria was always a gifted child. Eve Curie in her biography of her mother described some of the family dynamics. Maria's older sister Bronia enjoyed playing teacher with her little sister Maria, the student. Once when teacher Bronia was stumbling through a reading assignment, four-year-old Maria became impatient, grabbed the book, and read the opening sentence on the page perfectly. Bronia was humiliated, and Maria burst into tears, claiming that she "didn't do it on purpose." As Bronia sulked, Maria said that it wasn't her fault; "it's only because it was so easy!"[3] Instead of being delighted, her parents were concerned about her precociousness. They would have preferred that she was a little girl who played with blocks and dolls.

When Bronislawa became ill, much of her care devolved upon the oldest daughter, Zofia (known as Zosia). Her mother selected eleven-year-old Zosia to accompany her to the spa where she served her mother as a nurse, maid, laundry woman, and entertainer. Zosia was considered to be "delicate," and the family thought that bathing in the sulfur baths and taking long naps would restore her health. However, it seems that Zosia had little time for these healthful activities. Her constant attention to the needs of Bronislawa resulted in a closeness between the two and contrasted with her mother's behavior toward her youngest daughter, Maria, with whom she was physically distant. She did not seem to fret about exposing "delicate" Zosia to tuberculosis. Zosia attended a French school and was an excellent student. Bronislawa was proud of Zosia's school accomplishments and was overjoyed when Zosia announced that she was first in her class. Perhaps it was her background as a schoolteacher, but to Bronislawa Sklodowska it was very important that her children do well in school.

After two years at the spa, it became apparent that Bronislawa's health was not improving, so she and Zosia returned home. As if their mother's grave illness was not enough, politics intruded upon their lives causing Wladyslaw to lose his teaching position and position as under-inspector.

Under-inspector was the highest administrative position in the schools that a Pole could expect to hold. Wladyslaw's loyalty to Russia was always suspect, and when the Russian policy became increasingly more severe, he lost his job. The Russian government had embarked upon a policy known as Russianization, where not only was Poland's official language Russian but Polish officials such as Wladyslaw Sklodowski were replaced by Russian immigrants. The family lost its living quarters, status, and income. In order to survive, Wladyslaw turned the family's new home into a boarding school for boys. The house was raucous and overcrowded, and the school did not solve the Sklodowskis' financial worries as they had hoped. In fact, it may have had a more dire consequence for the family. Both Zosia and her younger sister Bronia contracted typhus in 1874. Often associated with wars and human disasters, typhus had existed in epidemic form in Poland ever since Napoleon's troops first invaded in the early nineteenth century.[4] Possibly the conditions at the crowded boarding school provided the source of the disease for Zosia and Bronia. Although Bronia recovered, fourteen-year-old Zosia, her mother's pride and joy, died. It took two more years before their mother finally succumbed to tuberculosis, but she never recovered from the death of the daughter who had given so much to her.

Although Maria, the baby of the family, had spent the least amount of time with their mother, the tragedy seemed to affect her more profoundly than the others. Bronislawa died on May 9, 1878, the day after she had called her children and husband into her bedroom to say goodbye. Ten-year-old Maria sobbed uncontrollably. In two years she had lost both a sister and her mother.

EDUCATION

Maria was only six years old when her father was forced out of his supplemental job as under-inspector. Her formal schooling had begun at the Freta Street School where her mother had been headmistress. In order to get to school Maria had to walk a long distance, so when she was in the third grade her parents enrolled her with her sister Helena in another private school closer to home. By this time, her mother's tuberculosis had gotten much worse, and the family took in a boarder, Antonina Tupalska, a math and history teacher who helped around the house and walked Maria and Helena to school. The children thought that bossy Miss Tupalska was hardly a suitable substitute for their mother. Once they arrived at the Madame Jadwiga Sikorska's private school, the girls received an excellent education. Both she and her sister Helena were in the same class in school, although Helena was a year older. It must have been difficult for Helena to find her younger sister surpassing her in class. Many years later, Helena recalled an incident in which Maria had forgotten to memorize a long passage in German. Since German was her third class, she used the two ten-minute breaks between classes to learn the passage. Helena complained that it had taken her several hours to learn the same passage. All of the Sklodowski children were fine students, but there is little doubt that Maria's special abilities caused some problems with her brother and sisters.

Although she was required to teach only in Russian, Madame Sikorska was able to conceal what she was really doing from the Russian authorities. While she was actually teaching the Polish language, geography, and history both teacher and students engaged in an elaborate cover-up. For example, home economics on the official program really stood for Polish history. Everyone in the school understood the deception. When the Russian inspectors came, students and teachers returned to the required course of study. During Maria's year at school the inspector was quite benign and seemed to empathize with the school's Polish sympathies. At one time he even warned them that the superintendent was coming the next day and that the children should not bring their Polish books to school. However, some of the subsequent inspectors were much more menacing.

Although Maria was the youngest student, she was also the brightest. Since she spoke Russian well she was often chosen to recite when the inspector came to visit. She wrote that "this was a great trial to me, because of my timidity; I wanted always to run away and hide." Maria later described her feelings as anger not timidity. I "wanted always to raise my little arms to shut the people away from me, and sometimes, I must confess, I wanted to raise them as a cat [raises] its paws, to scratch!"[5]

Maria's stern but kindly teacher Mlle. Tupalska was a patriotic Pole. She illegally taught the children Polish history in Polish. The entire class was trained to hide its books when a bell signaled the arrival of the Russian inspector. One morning when they were deeply engrossed in studying Polish history in the forbidden Polish language, the bell sounded. Immediately all remnants of Polish history vanished; the students were calmly sewing buttonholes in squares of material. The inspector asked the teacher to call on one of the students. The frightened victim was Maria. She was always selected because of the extent of her knowledge. After being satisfied that she answered all of his questions correctly, the inspector went off to another classroom. Maria was so upset after he left that she burst into tears.

From her father, Maria developed a love not just for science but also for literature, especially poetry. She easily memorized long passages from the great Polish poets and became proficient in French, German, Russian, and later English. Bronislawa was an accomplished musician, but Maria noted that although her mother had a beautiful voice, her own "musical studies have been very scarce."[6] She regretted the fact that when her mother was no longer available to encourage her, she abandoned her music. Her favorite subjects, however, were mathematics and physics. Her father encouraged her interests, but as she recalled in her autobiography, "unhappily, he had no laboratory and could not perform experiments."[7]

A dark cloud hovered over Maria after her mother's death. As Bronislawa became increasingly ill, she spent more and more of her time in church. Maria began to be a little jealous of a God who took so much of her mother's time away from the family. She also felt angry with this God who allowed bad things to happen to those she loved. Several years later, after her mother's death, Maria rejected religion completely.

After Bronislawa died, joy and laughter were very rare in the household. Maria's father mourned by becoming preoccupied with his work. The period of mourning lasted for several years, as was the custom in Poland. The windows had black curtains, women wore black veils, and notepaper was edged with black. The atmosphere was not a very healthy one for a sensitive young girl. Madam Sikorska realized that Maria was emotionally distraught and suggested to Wladyslaw that she stay out of school for a year. Her father rejected the kindly Sikorska's advice. He did pull her out of the nurturing environment of the private school, but he enrolled her in the government-run Advanced High School (Gymnasium) Number Three in downtown Warsaw in order to expose her to a more rigorous education. Even though Maria repeated the previous grade she was still younger than the other students. Prior to Russification this school had been German. German schools were well known not only for a cold academic environment but also for their appreciation of learning, and Maria was exposed to some excellent teachers. In spite of the fact that Maria received a superb education, especially in physics, Russian literature, and the German language, she at first despised the school. Maria and her friend Kazia found the superintendent of studies, Mlle. Mayer, particularly detestable. Their disrespect infuriated "Mayer" as they called her. Mayer was fixated on Maria's curly hair that refused to be confined into smooth braids. When reprimanded, Maria would look innocently at her teacher. This look further annoyed her teacher, who sputtered that Maria must not look down at her. Maria, who was a head taller than Mayer, replied that she could hardly do anything else. Maria and Kazia were discovered dancing in joy among the desks after the assassination of the Russian tsar, Alexander II, an incident that did not endear them to Mayer and the Russian authorities.

Maria did not want to admit that she was beginning to enjoy school. In a letter to Kazia during vacation, she admitted her guilty secret. "I like school. Perhaps you will make fun of me, but nevertheless I must tell you that I like it, and even that I love it."[8] Maria loved learning but took great delight in various pranks that she and Kazia played on their government teachers and students. Yet Maria was often under stress. Fear of the Russian inspectors, a compelling need to be the best in her class in all subjects, and continuing sadness over the deaths of her mother and sister added to her discomfort. She was on an emotional roller coaster vacillating between almost frenzied joy and deep depression.

As time went on, Maria found herself deeply involved in Poland's problems. Although one might expect that the oppressive conditions imposed upon the Poles by the Russians would stimulate another revolution, many Polish leaders decided that insurrection would not only be dangerous but would not work. Instead they decided to forego their goal of working for independence immediately and sought to supplant it with an attempt to strengthen the country through more subtle means, including education, economic development, and modernization. In other words, they wanted to fortify the country from the grass roots up. Many who supported this view considered it a stopgap measure while they awaited an eventual opportunity to become self-governing. Maria suffered along with Poland. But none of her personal hobgoblins interfered with her progress in school.

In 1883 she graduated from Gymnasium Number Three. Like her siblings Józef and Bronia, she finished first in her class and was awarded a gold medal. Fifteen-year-old Maria was younger than her fellow students. After graduation, the stress of her determination to be the best academically, the loss of her mother and sister, and the state of Poland caught up with her. As she wrote in her autobiography, "the fatigue of growth and study compelled me to take almost a year's rest in the country."[9] The country life at the estate of one of her two Boguski uncles without the pressure of responsibilities seemed to be the ideal cure.

Maria traveled south through the flat plains of Poland toward her uncles' houses. Although the first part of her journey was by train, the final part was over rutted roads in a horse and wagon. While bouncing along she was able to get a clear view of the Poland that she loved so well with its beauty smudged by the poverty of the people. She saw examples of peasant hardship with shoeless men and women plowing the fields. If they were fortunate they had oxen, if not they pushed the heavy plows by hand.

She first went to her mother's brothers, Henryk and Wladyslaw Boguski. Charming though they were, the two brothers were considered the black sheep of the family. Her other relatives criticized Uncle Henryk as a dilettante—a Jack of all trades and master of none who lived on his wife's income as the manager of the village store. Not very kindly, they described his wife as a simple woman. Uncle Wladyslaw, on the other hand, married a woman with a dowry but became involved with his brother in failed moneymaking schemes. Maria was not aware of the financial problems that her uncles were facing, and she found the atmosphere intoxicatingly merry. Interesting people came to visit, and the house was filled with books, music, and entertainment of all kinds.

The time with her uncles was carefree with little responsibility to work or study. Maria wrote to her school friend Kazia that "aside from an hour's French lesson with a little boy I don't do a thing, positively not a thing— for I have even abandoned the piece of embroidery that I had started." Sometimes she slept late or, if she felt like it, got up at four or five o'clock in the morning to walk in the woods; roll hoops; play battledore and shuttlecock, cross-tag, the game of Goose, and "many equally childish things." The books that she read were "only harmless and absurd little novels." During this time, Maria's intellectual interests lay dormant. "Sometimes I laugh all by myself, and I contemplate my state of total stupidity with genuine satisfaction."[10]

As summer passed, Maria traveled to the house of another uncle, this time her father's brother Zdzislaw. Uncle Zdzislaw and his wife, Aunt Maria Rogowska, lived in the foothills of the Carpathian Mountains with their three daughters. The lively group elevated Maria's spirits even higher. One of their most enjoyable activities was the *kulig*, a combination costume party, ball, musical, and sleigh ride. Maria wrote: "I have been to a *kulig*. You can't imagine how delightful it is, especially when the clothes are beautiful and the boys are well dressed." Sixteen-year-old Maria observed that "there were a great many young men from Cracow, very handsome boys who danced so well! It is altogether exceptional to find such good dancers." The party lasted all night, and "at eight o'clock in the morning we danced the last dance—a white mazurka."[11]

Maria returned to Warsaw, but in the spring she was able to extend her year of relaxation and fun. With her sister Helena, Maria was invited to spend the summer at the country estate of a former student of their mother. The manor house was beautifully appointed. Maria described it all to Kazia.

> I shall only say that it is marvelous. . . . There is plenty of water for swimming and boating, which delights me. I am learning to row—I am getting on quite well—and the bathing is ideal. We do everything that comes into our heads, we sleep sometimes at night and sometimes by day, we dance, and we run to such follies that sometimes we deserve to be locked up in an asylum for the insane.[12]

NOTES

1. In Polish, the family name of the man ends in an "i" and the woman "a." Thus Wladyslaw's (Maria's father's) family name is Sklowdowski, whereas Bronislawa's (Maria's mother's) is Sklowdowska. When speaking of the entire family, Sklowdowski would be used.

2. Not until 1882 did Robert Koch discover that tuberculosis was a contagious disease caused by a bacterium. An effective treatment was only developed many years later.

3. Eve Curie, *Madame Curie: A Biography* (Garden City, NY: Doubleday, Doran & Co., 1938), p. 9.

4. Typhus is caused by an organism, *Rickettsia prowazekii*, and is transmitted by the human body louse. The louse becomes infected when feeding on the blood of mammals that have acute typhus fever. Infected lice then excrete the microscopic rickettsia when feeding on another mammal. Humans or other animals are infected through scratching, resulting in rubbing louse fecal matter or crushed lice into a scratch, bite, or other type of wound.

5. Susan Quinn, *Marie Curie: A Life* (New York: Simon and Schuster, 1995), p. 45.

6. Marie Curie, "Autobiographical Notes," in *Pierre Curie,* trans. Charlotte and Vernon Kellogg (New York: Macmillan, 1923), pp. 160–61.

7. Marie Curie, "Autobiographical Notes," p. 161.

8. Eve Curie, *Madame Curie*, p. 36.

9. Marie Curie, "Autobiographical Notes," p. 163.
10. Eve Curie, *Madame Curie*, p. 40.
11. Ibid., p. 43.
12. Ibid., pp. 43–44.

Chapter 2

PREPARING FOR THE FUTURE

Eventually Maria had to return to reality. The year of rejuvenation had been exactly what she needed. But after her glorious vacation year, she had to return to dirty, dingy, depressing Warsaw and confront her future. Her choices seemed singularly unappetizing. Still, the time arrived when she had to decide what she was going to do with her life. A Polish girl at that time had only limited possibilities for higher education. The gymnasia for boys and girls differed in the subjects taught. The boys' gymnasia taught Russian, Latin, and Greek; the girls' gymnasia, on the other hand, did not teach classical languages. This difference might have been insignificant if it had not been for the entrance requirements for admission to the universities in the Russian empire, which at that time included Poland. These universities required classical languages, so that women were effectively blocked from admission. There was a solution for Maria, albeit an expensive one. She could leave her beloved Poland to study in a foreign country. The financial barrier seemed insurmountable. The family simply did not have enough money to educate the girls outside of Poland. Józef was attending medical school at Warsaw University and was in no position to help his sisters with their education. The alternative that Maria reluctantly decided upon was a teaching career in a girls' school. Thus, during her year back in Warsaw she tutored younger students while attempting to educate herself.

Their father challenged the Sklodowski children to learn. He was a very proper Polish gentleman who wore carefully brushed dark clothing and whose gestures and speech were always precise. He was a man who planned every aspect of life. If the family went on a holiday, the children could be certain that their father had organized every aspect of their trip in advance—what sights they were going to see, where they would stay, and how much they would spend. Where Wladyslaw Sklodowski was concerned, there was no spontaneity—nothing was left to chance. As a science teacher, he kept up with the progress in chemistry and physics. But he also knew Greek and Latin and was able to speak English, French, German, Russian, and (of course) Polish. He composed poetry and loved literature. On Saturday evenings the family would gather around Wladyslaw with a teapot steaming in the background and discuss literature. Living in this intellectual atmosphere gave Maria and her sisters an advantage unknown to most Polish girls. Forbidden official higher education, Maria vowed to educate herself.

Through her self-education program, Maria became involved in the Polish "Positivist" movement. Positivism was a philosophy that stressed the importance of scientific knowledge. It was begun by the nineteenth-century French thinker Auguste Comte (1798–1857). Positivists claimed that sensory experience was the most perfect form of knowledge. Maria and her friends, many of whom were students at Warsaw University, accepted this part of Positivism but modified it to suit their needs. They expanded its original intent to include ways of solving Poland's political and socioeconomic problems. The Positivist reform program was not revolutionary but gradual. It stressed the importance of trade, science, and industrial advances to Poland, areas which were considered to be beneath the Polish upper class.

There were two main reasons why Polish Positivism was so appealing to Maria and her sister Bronia. First was its stress on the importance of women. Although Comte himself was convinced of women's inferiority, in Poland the Positivists taught that women, if properly educated, could contribute to the reform movement. Therefore, in Warsaw many young Polish intellectual girls flocked around Positivist teachers in an informal setting. Maria admired the ideas of a Polish Positivist novelist, Eliza Orzeszkowa (1841–1910). Not only did Orzeszkowa write novels, she was also a part of the Polish literary scene espousing the notions of contemporary Positivist thinking. In her novels, she sought to educate her readers in order to change their attitudes and values. She hoped to eliminate class, race, and gender prejudices. Of course, she did not accept the part of Comte's Positivism that considered women naturally inferior. In her writings, Orzeszkowa discussed the education of the masses, the development of science, and class discrimination. They also reflected her strong belief in evolution and her agnosticism. All of these ideas became a part of Maria's intellectual persona for the rest of her life.

Positivism offered a second benefit to the patriotic Sklodowska girls who looked for ways to improve Poland's situation. The idea of industrial advances to replace the old romantic ideal of Poland appealed to them. Influenced by the evolutionary ideas of the English scientist Charles Darwin (1809–1882) as interpreted by her teachers and associates, Maria saw the salvation of Poland occurring through science and logic. Positivism was not just a passing interest for Maria. During her entire life she abhorred ideas that were unsupported by empirical evidence and insisted that education was necessary if social progress was to occur.

All of these ideas flourished underground, for Poland was still under the thumb of Russia. Maria and Bronia became a part of an institution known as the floating university where revolutionary ideas flowered. University was a far too pretentious name for this loosely constituted group of people. One writer described this as a "parochial little institution" consisting mostly of teenage girls as well as young married women with little else to do.[1] Another was more positive about this "little university," since it had a regular curriculum, with courses meeting for two hours a week.[2] In either case, this gathering kept young Polish women's interest in intellectual activities alive. It provided them with a forum where they could discuss new ideas such as Positivism and Marxism. The Marxist movement in Poland originated among the factory workers. Marxism and Positivism clashed on several issues. Whereas Positivists suggested gradual change and scientific solutions to Poland's problems, to Marxists the solution was more radical; they rejected collaboration with occupying powers and supported revolutionary change. Although she was interested in Marxist ideology, Maria remained a Positivist.

These ad hoc attempts to better themselves were not sufficient for Maria and Bronia. They both wanted desperately to obtain university degrees and plotted to find a way to get the money to study abroad. Bronia's dream was to go to the Paris Faculty of Medicine, which at the time was accepting many Polish women students, and get her MD degree. She then planned to return and open a medical practice in Poland. Her younger sister Maria was uncertain about what she wanted to study in Paris because she was interested in subject fields ranging from literature to physics. Maria's tutoring did not enable her to save enough money for her education. Since Warsaw was an expensive place to live, she realized that she would be an old lady before she saved enough money to go to a university.

EDUCATION OF WOMEN

Polish women were not the only European women who found it difficult to get a university education. Unlike in the United States where the women's colleges in the northeast—Vassar (1865), Smith (1875), Wellesley (1875), Radcliffe (1879), Bryn Mawr (1885), Barnard (1889), and Mount Holyoke (1893)—gave women the opportunity to obtain a university education, European women had no similar institutional support. England was also in the forefront of women's university education where, under the leadership of Emily Davies and Anne Clough, residential colleges for women were established. Girton (1869) and Newnham (1875) at Cambridge University, followed by the founding of Somerville (1879), Lady Margaret Hall (1879), St. Hugh's (1889), and St. Hilda's (1893) at Oxford, were among the early women's colleges. Although neither Oxford nor Cambridge granted degrees to women during the nineteenth century, the examinations at these universities were gradually opened to them. The provincial universities—Leeds, Manchester, Bristol, Durham, and Birmingham—were more hospitable to women students than Oxford and Cambridge. They followed the lead of the University of London (founded in 1836), whose charter stipulated the admission of women to the degree program without reservation. The situation was quite different at German universities. Throughout the nineteenth century women were unable to matriculate at German universities, although they had some access to these institutions by the end of the century: in 1891, the University of Heidelberg allowed women to attend as auditors; the University of Göttingen granted a PhD to the American physicist Margaret Maltby

(1860–1944) in 1895; the following year another American, the physiologist Ida Hyde (1857–1945), received a PhD from Heidelberg; and in 1899, the German physicist Elsa Neumann (1872–1902) earned a PhD degree from the University of Berlin. During the early twentieth century, the legal barriers to women's admission had for the most part crumbled, but because few German women had sufficient training to pass an entrance examination, most of the women who entered German universities were foreigners.

Although the situation varied from country to country, in the late nineteenth century the education of women in Europe advanced more than it had previously. In France 109 academic degrees were conferred upon women between 1866 and 1882. Switzerland, Sweden, and Denmark all opened their universities to women in the third quarter of the century. Although many Italian universities had accepted women students and faculty members during the Middle Ages and the Renaissance, they had closed their doors to women during the late eighteenth and early nineteenth centuries. However, they began to readmit women in the 1870s.

Maria Sklodowska's situation was very similar to that in Russia because, of course, the Russians were in charge of her part of Poland. In Russia, the government rejected the petition of women to be admitted to the universities in 1867. These women participated in an informal system of education whereby cooperating professors, by a combination of public lectures and discussion meetings in private homes, were able to present a complete course. The Russian mathematician Sofia Kovalevskaia (1850–1891) was a product of this system of underground education. Maria's floating university followed this pattern. Unlike Kovalevskaia, she did not choose a common path to the university favored by Russian women. Since women who wanted to leave Russia to attend a foreign university could not do so unless they were married, a woman often found a young man who would agree to a "sham" marriage. Maria's approach was much more conventional. There were no laws in Poland to prevent her from leaving the country, so when she earned the required sum she would be able to leave legally. However, she could not accumulate the necessary money by tutoring.

Maria's father was unable to educate his daughters because he had engaged in a speculative financial venture in which he lost the nest egg that he had accumulated. He lamented that he was unable to send the girls abroad and give them the brilliant educations that they deserved. He feared that not only was he unable to help them but he might become dependent on them for his own survival. Describing the difficulties of her predicament Maria wrote, "I resolved to accept a position as governess." Her first job was a disaster. She took a position as governess for the family of a lawyer. Vehemently expressing her anger in a letter to her cousin, she complained that the family members spent money foolishly on luxuries, and were then too cheap to buy oil for the lamps. "They pose as liberals," she ranted, while "in reality, they are sunk in the darkest stupidity."[3] She described them all as terrible gossips. Maria had no desire to stay in such a household, particularly since she and the lady of the house were mutually hostile.

With both Bronia's and her own goals in mind, Maria developed a plan whereby each sister would get the education that she so wanted. Maria would find another position as a governess, this time outside of Warsaw, live on a pittance, and save the majority of her salary. Bronia, as the oldest, would be the first to benefit from this plan. When Bronia finished medical school, she could, in turn, help Maria. Bronia had saved enough money to pay for the trip to Paris and to support herself for a year. She decided to begin her medical training in Paris right away. Maria as a governess "with board, lodging and laundry all free," would have "four hundred rubles a year in wages, perhaps more," to contribute to Bronia's subsequent years of schooling.[4] In spite of regrets for Maria's sacrifice, Bronia agreed that the plan might work. While it was easy to be altruistic in the abstract, Maria found the reality of keeping her promise to Bronia very difficult. The five years that it would take for Bronia to finish medical school seemed interminably long.

Maria's second try as a governess was much more successful, although taking the well-paid position meant that she would be away from her family and her beloved Warsaw. With her face pressed against the train window, she watched familiar landmarks recede into the distance. After traveling for several hours by train she still had to drive five hours more by horse and sleigh before she reached her destination. She initially wrote glowing reports about her new employers, the Zorawskis. In a letter to her cousin Henrika Michalowska on February

3, 1886, she described them as "excellent people." Besides the parents, the Zorawski family consisted of an older daughter "about my age" and "two younger children a boy and a girl."[5] They also had three sons who were being educated in Warsaw. She worked with Bronka, the older daughter, for three hours each day and with Andzia, the ten-year-old, for four. She wrote Henrika complaining that "Andzia, who will soon be ten, . . . is an obedient child, but very disorderly and spoiled."[6] As time went on, Maria demonstrated that she found coping with this child grated on her nerves. She no longer spoke of Andzia as obedient and described to Henrika how angry she became when Andzia did not obey her.

> Today we had another scene because she did not want to get up at the usual hour. In the end I was obliged to take her calmly by the hand and pull her out of bed. I was boiling inside. You can't imagine what such little things do to me: such a piece of nonsense can make me ill for several hours. But I had to get the better of her.[7]

Although she spent over seven hours a day tutoring the Zorawski children and another hour with the son of one of the Zorawskis' servants whom she was preparing for school, she did have some free time. With the blessings of the Zorawskis, she used this time to teach young peasant children how to read and write. Her pupils were peasants from the Zorawskis' beet farms and the workers' children from the sugar beet factories. This activity took not only commitment but courage on Marie's part. The Russian government would not approve of such an activity and it was even more dangerous because she circulated Polish books to the children's parents. If she had been caught the possible punishment was imprisonment or deportation to Siberia. Her first class consisted of ten children and the number soon grew to eighteen. She taught them in her own room.

In spite of her seven official hours of work a day and many hours tutoring the peasant children, Maria was often bored. She found most of the people her age that she met to be shallow. They seldom thought about the social, philosophical, and economic problems that so obsessed her and stimulated her to teach the peasant children. The exception was eighteen-year-old Bronka. Due to Maria's superior knowledge, she was Bronka's teacher. The two became good friends even though Maria as a governess was considered by the family to be inferior to Bronka in social class. Social class for a governess presented a contradiction. In order to be a governess, a person had to be "well-born," well educated, and have impeccable manners. However, the position was not considered one of high social class.

The question of class muddied the waters in her relationship with Bronka and later with the older Zorawski son, Kazimierz, who was studying mathematics at the University of Warsaw. On one of his holidays at home he met Maria. The two promptly fell in love. Kazimierz's parents immediately attempted to break up the romance, for they considered Maria merely an employee—a lowly governess without money or status. It did not matter to the Zorawskis that Maria was intelligent, came from a good family, and was obviously a refined person. His parents' views came as a shock to Kazimierz, who had expected that they would approve of their engagement. The very opposite occurred. His father fell into a rage. His mother almost fainted. They could not imagine that their golden boy, who could have married any girl who he wanted, had chosen one of their employees. Although Kazimierz resisted his family for a while, he finally went along with his parents' wishes. Brokenhearted Maria swore that she would never marry nor fall in love again. Although she was angry with the Zorawskis, she continued as governess for fifteen more months. During this time she wrote vitriolic and desperately miserable letters to her friends and family at home. Other people's good fortune was especially hard for her to take when she herself was so unhappy. After receiving a cheerful letter from her friend Kazia, Maria replied stating that she had been disconsolate and did not want to hear about Kazia's happiness.

The years at the Zorawskis had benefits as well. Maria embarked upon a program of self-education, starting her studies at nine o'clock in the evening and getting up at six in the morning. At one point she was reading books in physics, sociology, and anatomy and physiology. She explained that she preferred to read several books at a time rather than just concentrating on a single subject. When reading became too tedious, she worked on problems in algebra and trigonometry. Even though she found literature and sociology as interesting as science, she eventually decided that her future lay in mathematics and physics.

In 1889 it looked as if she could leave her job with the Zorawskis once and for all. Perhaps she would finally have a chance to go to Paris. Her sister Bronia had become engaged to a fellow medical student, another Kazimierz—Kazimierz Dluski. After they married, Maria might be able to go to Paris to attend the university. Until then, she was content to stay in Warsaw with her father. She soon took another position as a governess with the Fuchs family. Although it was less humiliating than her previous experience, it was still very depressing. When the letter from Bronia finally came to invite Maria to come to Paris, she replied with a litany of reasons why she could not go—her father was too old and would be disappointed, she must help her sister Hela find a position in Warsaw, and many other excuses. At the end of the letter she did not mention Paris further, but clearly was ambivalent about her sister's marriage when she wrote, "My heart is so black, so sad, that I feel how wrong I am to speak of all this to you and to poison your happiness, for you are the only one of us all who has had what they call luck. Forgive me, but, you see, so many things hurt me that it is hard for me to finish this letter gaily."[8] Marie stayed with her father for the year. For the previous two years, Wladyslaw had directed a correctional agricultural colony outside of Warsaw, a position that he despised. But when Maria returned he had retired, and father and daughter were able to live comfortably in his apartment. During this year, she had access through her cousin Józef Boguski to a laboratory for the first time in her life. She described it in her autobiography as a small, municipal physical laboratory. Although her time was usually confined to evenings and Sundays, she was left alone to try out the experiments that she had previously only read about in textbooks. When her laboratory experience was not successful, she was plunged into despair. When the experiments succeeded, she was elated. This experience, she explained, confirmed her interest in the fields of experimental physics and chemistry.

NOTES

1. Robert Reid, *Marie Curie* (London: Collins, 1974), pp. 24–25.

2. Susan Quinn, *Marie Curie: A Life* (New York: Simon and Schuster, 1995), pp. 65–66.

3. Reid, *Marie Curie*, p. 29.

4. Eve Curie, *Madame Curie: A Biography* (Garden City, NY: Doubleday, Doran & Co., 1938), p. 58.

5. Marie Curie, "Autobiographical Notes," in *Pierre Curie*, trans. Charlotte and Vernon Kellogg (New York: Macmillan, 1923), p. 164.

6. Reid, *Marie Curie*, p. 30.

7. Ibid., pp. 31–32.

8. Eve Curie, *Madame Curie,* pp. 84–85.

Chapter 3

PARIS AND THE SORBONNE

It was not just her responsibilities toward her father that kept Maria in Warsaw. She seems to have attempted to reconcile with Kazimierz one last time. Although information is not available about what happened when they met that summer, the relationship was over. Maria wrote another letter to Bronia asking if she could still come to Paris. Bronia wrote back and extended the invitation. This time Maria did not hesitate. She boarded the Paris train and sat on a folding chair surrounded by her luggage in the fourth-class carriage. After three miserable days, her train chugged into the Gare du Nord (one of the Paris train stations) where she was met by her brother-in-law, Kazimierz Dluski (Bronia was visiting in Poland), who took her to their apartment. Kazimierz wrote to her father saying, "Everything is going very well with us. Mademoiselle Marie is working seriously; she passes nearly all her time at the Sorbonne and we meet only at the evening meal."[1] He showed some displeasure with Maria's self-reliance. "She is a very independent young person, and in spite of the formal power of attorney by which you placed her under my protection, she not only shows me no respect or obedience, but does not care about my authority and my seriousness at all."[2] Apparently, Marie (as she was now called taking the French form of her name) felt smothered by her overly intrusive brother-in-law. Kazimierz wanted to have people around him at all times. Whereas Bronia was an able and willing hostess, Marie was unwilling to take over these duties while Bronia was in Poland or participate in them after she returned. She also was expected to go out in the evenings to the theater and to concerts. The Dluski house was often filled with interesting people such as musicians, scientists, and Polish political activists. Although Marie enjoyed going to concerts and meeting the Dluski guests, she disliked the time they took from her studies.

Kazimierz began to get on Marie's nerves. Although he wrote to her father that he and Marie understood each other well and "lived in the most perfect agreement," Marie complained to her brother, Józef, that her "little brother-in-law" disturbed her constantly and that he insisted that she do nothing but "engage in agreeable chatter with him."[3] To one for whom studying was her very reason for living—she had dreamed her whole of life of learning in Paris—it was intolerable that she would be kept from concentrating by Kazimierz's incessant conversation. The Dluskis' residence was even more chaotic because Kazimierz, now a practicing physician, saw his patients in the house. Using the excuse that she needed to be closer to the university, Marie found herself an attic room and moved away from her family.

Far from being lonely, Marie relished living by herself. She could have the freedom to do exactly what she wanted to do when she wanted to do it. Money, however, was a constant problem. She had given up free room and board and now had to support herself on her meager savings and the small sums her father sent her. Living close to the university in the Latin Quarter she lived in poverty, as did other students. Marie's status was equally wretched. Her attic room (she was at the top of six flights of stairs) was without heat, lighting, and water. She had to fetch water from the landing in her pitcher even to make a cup of tea. Her furniture consisted of a mattress she had brought from Poland, an iron folding bed, a white wooden table, and a kitchen chair. She had a stove for heating and a petroleum oil lamp for reading at night. Marie, who had never learned to cook, was ignorant of how to prepare even the simplest meals. She was too poor to eat in the Parisian restaurants. She nearly starved during the first months by herself. She boiled an occasional egg on an alcohol burner and sometimes had a piece of chocolate or fruit. Her standard diet was buttered bread and tea. Not surprisingly, she became ill and often fainted. After Kazimierz discovered her situation, Bronia and he brought her back to their apartment, gave her medicine and healthy food, and she soon became strong again. She did not keep her promise to treat herself better, for as soon as she returned to her attic, she went back to her former ways. Panicked about the impending examinations, she, as her daughter Eve later wrote, "began again to live on air."[4]

It was not surprising that Marie was so concerned about her examinations. When she entered the university she soon realized that she was not sufficiently prepared to follow the physical science course. French students spent at least seven years preparing to enter the university. Although Marie had equipped herself as well as possible, she soon understood that she had to work doubly hard to eliminate her deficiencies, particularly in mathematics. In her autobiographical notes she wrote:

> I divided my time between courses, experimental work, and study in the library. In the evening I worked in my room, sometimes very late into the night. All that I saw and learned that was new delighted me. It was like a new world opened to me, the world of science, which I was at last permitted to know in all liberty.[5]

As Marie knew, the Sorbonne was one of the oldest universities in the world. By 1253 it had a recognized Theology Faculty and in 1271 it also became a Faculty of Philosophy and Arts. Following the Franco-Prussian War in 1870 and the events of the Paris Commune, French universities no longer provided the best education in the sciences. The Ministry of Education had invested little money in laboratories, so that research was no longer on the cutting edge. The French saw the need to reform their universities after France's catastrophic defeat by Prussia. Blaming the loss in large part on their inferior educational system, they began to reexamine their educational institutions. Even as they scorned their own universities, they heaped praise on those German universities that financed research laboratories and instituted seminars oriented toward research topics and methods. The French system was based on eloquent lectures and carefully argued theses, whereas the German professors discussed current research and trained students in practical laboratory work.

By the time that Marie Curie attended the university, improvements had been made in the science curriculum, although the German universities continued to outperform the French. Still, much of the reform of the Sorbonne had already occurred. Although the Sorbonne had once been a bulwark of church doctrine, the reformed institution preached republican anticlerical teachings. This emphasis appealed to Marie, who championed the superiority of the rational over the irrational. Theology was banished and the humanities de-emphasized with the sciences gaining the upper hand. The Sorbonne also was undergoing a massive building project at this time with science classrooms and laboratories being constructed.

More important to Marie's education than the general reform philosophy and the quality of the buildings, were the teachers. She found the French science courses challenging and many of her teachers inspiring. These teachers included the Nobel Prize winner in physics Gabriel Lippmann (1845–1921), who brought the German laboratory perspective to Paris. Talented in designing instruments, he emphasized the practical applications of physics. His sensitive devices were used in seismology and astronomy, and he received the Nobel Prize in 1908 for devising a method of photographic color reproduction. She also received training from Joseph Boussinesq (1842–1929), a physicist of the old school, who remained opposed to relativity theory and its consequences, but taught her the details of classical physics (the physics of Sir Isaac Newton [1642–1727]). In contrast to Boussinesq's practical and experimental emphasis, she was also taught by one of the most brilliant theoretical physicist/mathematicians of the time, Henri Poincaré (1854–1912), who made many novel contributions to mathematical theory and to celestial mechanics. Philosophically, he came close to developing a relativity theory himself. Thus, Marie was exposed to a variety of ideas in physics from these professors as well as others among the sixteen who taught her during her career at the Sorbonne.

Nearly all of the students in her classes were men, and the few women who were there were foreigners like herself. Constraints resulting from French ideas that boys and girls should have different types of secondary school education meant that Frenchwomen were less likely to enter their own universities than foreign women. In all fields of study, foreign women outnumbered French women at the Sorbonne until 1912 when the numbers of Frenchwomen finally exceeded those of foreign women. If nineteenth-century French women were educated at all, their education was mediocre. The daughters of the rich could attend private schools, some of which had been established at the end of the seventeenth century by Madame de Maintenon (1636–1719) and Bishop François Fénelon (1651–1715) to encourage the education of girls. However, education for the majority of French children, both boys and girls, only became possible when the government provided free schools. The education of girls continued to lag behind that of boys, because it was assumed that the girls would be taught at home by their mothers. The first government-funded school for young girls was created in 1807, with the purpose of educating the close relatives of members of the Legion of Honor. Although the education that these rich young women received included religion, reading, spelling, botany, some history and geography, and the art of being agreeable, it seldom was of a very high quality.

French secondary-school education was first established in 1867, although the promising beginning did not last. However, the situation did improve after the Camille Sée law—which required establishing secondary schools for girls—was passed on December 21, 1880. Seventy-two lycées and colleges were authorized with a more rigorous course content than the earlier schools. These schools were still inferior to those available to young men and did not prepare women for the baccalauréate examination required for entrance to the university.

With these strictures, it is not surprising that the number of foreign women students surpassed French students at the University of Paris. Foreign students were treated differently than French students and had a good deal more freedom in their conduct. The French theorist Jules Michelet (1789–1874) reported that the worst fate for a woman was to live alone. If a French woman (presumably including university students) would go out in the evening she would be taken for a prostitute. "If she were late, far from home, and became hungry, she would not dare enter a restaurant. . . She would make a spectacle of herself."[6] As a foreign student, Marie had much more freedom. Nevertheless, according to an American woman who went to France in 1900, a single woman could now discretely attend the theaters if "she is quiet in her dress, and is careful not to loiter in the foyers. People in Paris have begun to discriminate between two kinds of lone ladies."[7] It was unlikely that studious Marie would be attending the theater. However, as a foreign student at the Sorbonne, she was expected to have different rules of conduct. She also came from a family where women led independent lives. Even though Poland's women were denied access to higher education, they were often outspoken. Still, French people often made fun of the foreign students. One chronicler, Henri d'Almeras, ridiculed the foreign female students as working "with great patience, as though they were doing embroidery,"[8] He continued by saying that study made them ugly and that they usually wear glasses and look like school teachers.

Marie gradually came out of her shell and found that some of her fellow students wanted to be friendly. However, most of their interactions were concerned with studies. She also made friends with Polish students, none of whom were in the physical sciences. Two of the students were mathematicians and one was a biologist (he later married Marie's sister Helena), and a future president of the Polish Republic. She joined them in walks, political talks in bare rooms, and general reminiscences about home. They prepared Polish food for Christmas and organized theatrical performances. Although Marie did not have the leisure to learn parts for the plays, she did participate in the performances in other ways, much to the chagrin of her father when he found out. In a letter to Marie he wrote that he deplored her taking part in the theater. "Even though it be a thing done in all innocence, it attracts attention to its organizers, and you know that there are persons in Paris who inspect your behavior with the greatest care." He warned her that when she returned to Poland she might be in trouble if her name was mentioned in reference to events such as concerts, balls, and the theater. Such publicity might keep her from obtaining a job in certain professions. He concluded by remarking that "it would be a great grief to me if your name were mentioned one day."[9]

It is unclear how much her father's disapproval affected Marie's behavior. However she reported that after her first year she had to give up these relationships in order to devote all of her energy to her studies. "I was even obliged to devote most of my vacation time to mathematics."[10] As the time for her to take her first exam in physics drew frighteningly near, she became more and more withdrawn. She threw her entire being into her studies and resented any intrusion that would take her away from them. Not only did Marie have to master physics for her exams, she also had to take them in a language with which she was not completely at home. Her self-confidence would plummet if she did not do well, and she fretted about how she would perform on the examinations. However, looking back on those years of intense study she later characterized them as the best times of her life. The concentrated study paid off, and in 1893 she not only passed the *licence* exam in physics (a step beyond a bachelor of science degree) but also was first in her class. When her father heard that she was taking her examinations, he was overjoyed because he was certain that Marie would come home to him. In a letter to Bronia, he wrote that he intended to keep the lodging that he now occupied for himself and Maria when she returned. For her part, Marie realized how indispensable a mathematics background was to physics and chemistry and decided to return the next year and work on an additional degree in mathematics. Because she had been so successful as a student she received a six-hundred-ruble Alexandrovitch scholarship for her studies in Paris. On her return to Paris, she wrote to her brother, Józef, on September 15, 1893, explaining that she was

> studying mathematics unceasingly, so as to be up to date when the courses begin. I have three mornings a week taken by lessons with one of my French comrades who is preparing for the examination I have just passed. Tell Father that I am getting used to this work, that it does not tire me as much as before, and that I do not intend to abandon it.[11]

At the end of the next year she emerged with a *licence* in mathematics. She ranked second in her examinations.

Writing of these years, Marie noted that although they were sometimes painful, they also "had a real charm for me."[12] Being unknown in Paris was neither frightening nor particularly lonely to her. The feeling of independence made up for any anxiety she may have felt from being in a strange country, speaking a foreign language, and largely without friends.

NOTES

1. Eve Curie, *Madame Curie: A Biography* (Garden City, NY: Doubleday, Doran & Co., 1938), p. 98.
2. Ibid.
3. Susan Quinn, *Marie Curie: A Life* (New York: Simon and Schuster, 1995), pp. 45, 88.

4. Eve Curie, *Madame Curie*, p. 110.
5. Marie Curie, "Autobiographical Notes," in *Pierre Curie*, trans. Charlotte and Vernon Kellogg (New York: Macmillan, 1923), p. 171.
6. Claire Goldberg Moses, *French Feminism in the Nineteenth Century* (Albany: State University of New York Press, 1984), p. 35.
7. Mary Abbot, *A Woman's Paris: A Handbook of Every-day Living in the French Capital* (Boston: Small, Maynard & Co., 1900).
8. Quoted in Quinn, *Marie Curie: A Life*, p. 95.
9. Eve Curie, *Madame Curie*, p. 103.
10. Ibid., p. 172.
11. Ibid., p. 115.
12. Marie Curie, "Autobiographical Notes," p. 171.

Chapter 4

PIERRE AND MARIE

After her disastrous relationship with Kazimierz Zorawski, Marie had no time for romance. Declaring that she would never marry, she planned to devote herself full time to her studies. She had an ardent admirer, a Monsieur Lamotte. We only know of their relationship through his farewell letter to her. As she prepared to take her last exams in June 1894, she received a letter from him in which he promised not to disturb her by saying good-bye in person. Wishing her happiness and success, he wrote "one small word still of reproach: you insisted that I would quickly forget you when I had lost sight of you." This, he insisted was a mistake—he would always remember her. Although "without doubt we won't meet again . . . if you ever should need it, remember that you have left somewhere a friend ready to do everything possible for you. Adieu!"[1]

During the time that Marie was finishing her mathematics degree and was seeing Lamotte, she was hired by an organization formed to promote French science, the Society for the Encouragement of National Industry. Her tasks included a study of the magnetic properties of various steels, but she was severely limited by the lack of laboratory space in which to work. While she searched for adequate space, a Polish physicist (Jozef Kowalski) and his wife whom she had met during her days as a governess were in Paris for their honeymoon. After hearing of Marie's need, Professor Kowalski suggested a meeting with his friend Pierre Curie (1859–1906), who was working on magnetism at a nearby institution and might have space available. The Kowalskis may have hoped that a romance would result from the meeting. If this was true, their hopes succeeded far beyond their dreams. Pierre was thirty-four years old and a professor at the École de Physique et Chimie Industrielles (School of Industrial Physics and Chemistry) in Paris when they met. He and his brother Jacques discovered the phenomenon of piezoelectricity from their collaborative research.

Neither Pierre nor Marie had any idea of an impending romance. Pierre had vowed to live like a monk, after a young woman with whom he had been in love died. And Marie, after she had completed her two examinations, planned to return to her father in Poland, find a job, and use her new skills as a teacher and for the political betterment of Poland. She certainly convinced her suitor, Lamotte, that she would never return to Paris. The first night that Marie and Pierre met they found that, in spite of their differences in background, they had many common interests. They were immediately attracted to one another and by the time Marie left for Poland, Pierre was convinced that he wanted to marry her.

Pierre was the second son of Sophie-Claire Depouilly and Dr. Eugène Curie. Both sides of his family boasted scientists and inventors. Sophie-Claire's father and brothers were commercial inventors, and both Eugène and his father, Paul, were physicians. The Curies advocated revolutionary ideas in politics, religion, and science. Pierre absorbed the idealistic views of his ancestors and, although he was interested in social questions, he largely had put them aside for scientific ones. Like Marie, Pierre was shy and introverted and totally devoted to science. They shared a distrust of traditional religion.

Since Eugène Curie was an avid republican and skeptic, neither of his sons was baptized nor exposed to religion. Nevertheless, they were imbued with a sense of reverence for their environment. Pierre was of a contemplative nature, and his father, Eugène, decided that a traditional school would have a detrimental effect on his retiring, daydreaming son. Thus, he decided to educate both sons at home. Jacques was convinced that Pierre never got the well-rounded education that would have prepared him for a traditional university career. Although Pierre was precocious in science and mathematics, his education in literature and the classics was meager. It was fortunate that Pierre's parents recognized that he had an unusual type of intelligence. A dreamer, Pierre would have been considered a slow learner in school. With the freedom allowed by his untraditional education, Pierre learned to appreciate natural phenomena. Walking in the woods around Paris, he was able to clear his mind of extraneous thoughts, and look for intricate patterns in nature. In an entry in his diary from 1879, he praised his time in the country, reporting, "what a good time I have passed there in that gracious solitude, so far from the thousand little worrying things that torment me in Paris."[2]

His progress in mathematics and physics allowed him to earn his secondary school degree when he was sixteen years old. From an early age he was convinced that he wanted to be a laboratory physicist. After earning his undergraduate degree, he studied for the licentiate in physics at the Sorbonne, part of the University of Paris. He received this degree in physical sciences when he was eighteen years old. Because of excellent recommendations from the director and the assistant director of the laboratory where he worked on his licentiate, he was appointed preparator for the director and was put in charge of the students' laboratory work in physics, a position that he held for five years. During this time he began his experimental research. Because he needed money to support himself, Pierre was unable to continue his formal studies for the two or three years necessary for the doctorate.

Until he met Marie, Pierre had spent much of his life with his older brother Jacques. The brothers began to work together in physics when Pierre was twenty-one and Jacques twenty-four. From very early times, observers had noted that certain kinds of crystals placed in a fire attracted particles of wood and ash to their surfaces. In nineteenth-century France, the study of these crystals became popular. Tourmaline, for example, was characterized by having different crystalline faces acquire different electrical charges when heated to different temperatures. This phenomenon of generating small amounts of electricity was known as *pyroelectricity* (electricity from fire). Jacques and Pierre postulated that this phenomenon was not caused by heat but by pressure. When pressure was applied, opposite faces of the crystal should acquire an electrical charge. They devised instruments and experiments to test their theory that mechanical energy could be converted into electrical energy. The phenomenon was later named *piezoelectricity* from the Greek, meaning "to press." The instrument that they used to investigate many different crystals was called an electrometer, and it provided a method of measuring small electric currents. From 1880 to 1882 the brothers published seven papers on this topic.

To Pierre, collaboration seemed to be the normal way to do science.

He had begun to collaborate with Jacques from their late teenaged years at the Sorbonne where both were laboratory assistants. According to Paul Langevin, one of Pierre's early students, Pierre required a laboratory close to those he loved. Jacques and Pierre worked together until Jacques married and left Paris for Montpellier where he had a university appointment. Thereafter, their collaboration was confined to the summer months. After Jacques moved, Pierre left the Sorbonne for a job as laboratory chief at the School of Industrial Physics and Chemistry. The new position was certainly not a promotion. The buildings were old, his experimental research had to be put on hold, and he had only one assistant in his laboratory. But these obstacles that would have proved insurmountable to an ambitious researcher did not bother Pierre at all. He loved his students, was excited about teaching, and was absolutely content to remain on the same rung of the teaching ladder without a promotion. Many political games had to be played in order to be promoted in a university. To be pleasant to superiors, to attend social functions, and to network with those in power did not appeal to Pierre and they were all necessary activities if he was to be promoted. Freedom was also very important to Pierre. He was grateful to the director of the school, writing that he "allowed us all great liberty; his direction made itself felt chiefly through his inspiring love of science. The professors have created a kindly and stimulating atmosphere that has been extremely helpful to me."[3] When advised that a physicist was planning to resign and that he should become a candidate for the position, he replied, "it is a nasty job being a candidate for any place at all, and I am not accustomed to this sort of exercise." When he was proposed for an award by the director of the school, he refused, writing to the director that

> Mr. Muzet has told me that you intend to propose me to the Prefect again for decoration. I write to beg you to do no such thing. If you obtain this distinction for me, you will put me under the obligation of refusing it, for I have quite decided never to accept any decoration of any sort.[4]

As Marie mentioned in her biography of Pierre, his appointment at this school was a disaster for his experimental research. The partitions in his laboratory were not even in place. He had to practically build his entire laboratory from scratch. Because he was forced to interrupt his experimental plans, he became engrossed in theoretical research on crystals. He published a number of papers on the symmetry of crystals. This study was very abstract and involved Pierre's love for mathematics and focused on thought rather than experimentation.

Many years before he met Marie, Pierre wrote in his diary that

> women of genius are rare. And when, pushed by some mystic love, we wish to enter into a life opposed to nature, when we give all our thoughts to some work which removes us from those immediately about us, it is with women that we have to struggle, and the struggle is nearly always an unequal one.[5]

Pierre found Marie the rare woman of genius. She was fascinated by his scientific ideas and he was proud of her success in her examinations. The two seemed to complement each other—Marie was determined and focused while Pierre was imaginative and a dreamer. Both were idealists, but they acted out their idealism in different ways. Marie's idealism demanded that she return to Poland to contribute toward Poland's national spirit. Pierre's hopes were pinned on Marie's returning to Paris to marry him.

Pierre was convinced that science was the only certain path to rectify social injustices since social movements often failed and even successful social reformers sometimes did more harm than good. In his letters, he pushed hard to convince Marie that together they could use science for the good of human kind. Although the idea of a collaborative partnership was familiar to Pierre because of his close collaboration with his brother, Marie, on the other hand, had always worked alone and the way in which she would have planned to "save the world" would have been through active political participation. Yet she was eventually won over to Pierre's ideas. After having been burned by her unfortunate experience with Kazimierz Zorawski, she was extremely reluctant to allow herself to fall in love again. But the tall, lean, kindly Pierre, with the soft expressive eyes, with whom she shared so many interests and values, began to convince her.

But winning her over was not easy for Pierre. Although he recognized how important independence was for Marie, he wrote in exasperation, "I find that you are a little pretentious when you say that you are perfectly free. We are all more or less slaves of our affections, slaves of the prejudices of those we love."[6] Apparently Paris held a great attraction for Marie in the person of Pierre. During the summer that she was in Poland, Pierre begged her to return, writing that it would be a "beautiful thing, a thing I dare not hope, if we could spend our life near each other hypnotized by our dreams: your patriotic dream, our humanitarian dream and our scientific dream."[7] While she was in Poland, they wrote many letters back and forth, he attempting to convince her to return and she writing of practical concerns. He explained that there would be many career possibilities in Paris if she were French. And, if she married a Frenchman she would be considered French.

Marie did return to Paris, but did not accept Pierre's suggestion that they rent an apartment together. He had written that he had found one on Rue Mouffetard with windows overlooking a garden and which was divided into two separate and independent parts. Instead, she took an apartment next to Bronia's newly established medical office. Still, when Marie announced that she planned to live in Poland, Pierre was desperate enough to vow to go there with her. He promised that he would make a living in any way that he could, even by giving French lessons. To help convince Marie, Pierre enlisted the help of Bronia, whom he had already persuaded to assist him. Each would have to modify his or her principles in order to marry.

When Pierre decided to marry Marie he was strapped for money. He needed enough money for his work and for his forthcoming marriage. In order to do this he felt that he compromised his ideals by accepting a position as technical adviser to a Parisian optical firm at a fee of one hundred francs a month. He also would receive royalties by allowing the firm to exploit a photographic lens he had devised. Still, the amount of money that he received would not be enough to provide them with a comfortable lifestyle. If Marie had been the kind of person for whom money was important she never would have agreed to marry Pierre. He would have been a poor choice for a husband—someone who might seem to others to lack ambition. A kind and loving man who shared her interest in science and social rectitude was the kind of person Marie found attractive. Nevertheless, this stubborn woman was not easily won over by the equally determined Pierre.

Clearly, Marie's relationship with Pierre was heating up several degrees, for after she returned to Paris he convinced her to meet his parents. They lived in an aged small house at Sceaux in a Parisian superb. The house was surrounded by a lush garden. Marie found crusty old Eugène Curie delightful. A tall man with brilliant blue eyes, he impressed Marie with his sharp intellect. She found his passion for the natural sciences particularly alluring. As much as he loved the sciences, he had been unable to pursue them professionally for marriage and his family took precedence, and he was forced to practice medicine in order to support his family. He instilled in his sons the same love for science as he himself had and supported them in their scientific aspirations. Although Marie characterized Eugène as authoritarian, she also noted that he was unselfish, loving, and helpful to others. Pierre's mother was slight in build, and her health had been uncertain since the birth of her sons. However, she made the Curie home attractive and gracious.

Shortly after meeting his parents, Pierre invited Marie to the Sorbonne as a guest at the public examination of his doctoral thesis on magnetism. Marie was suitably impressed by his answers and realized even more acutely that they shared the same interests and values. In spite of their shared ideals, Marie was much more ambitious and was more willing to profit by the results of science. Pierre, on the other hand, rejected academic prizes. His self-effacing manner kept him from obtaining what he really wanted, a professorship. Although Marie cared very little for material things, she was more able to see the advantage of some kind of reward for achievements. After Pierre presented his thesis at the Sorbonne in March 1895, a professorship was created for him at the École de Physique et Chimie.

In her biography of Pierre, Marie states very matter of factly that

> after my return from my vacation our friendship grew more and more precious to us; each realized that he or she could find no better life companion. We decided, therefore, to marry, and the ceremony took place in July, 1895. In conformity with our mutual wish it was the simplest service possible,—a civil ceremony, for Pierre Curie professed no religion, and I myself did not practice any.[8]

If it were not for her brother-in-law's mother, who gave Marie a wedding dress, she might have worn the one dress that she owned to her wedding. Practical Marie scorned the idea of a white dress, saying to Kazimierz's mother that if she was going to be kind enough to provide a wedding dress, "please let it be practical and dark, so that I can put it on afterwards to go to the laboratory."[9] Bronia was more fashion conscious than Marie, and guided her sister to a dressmaker who made her a tasteful navy-blue wool suit and a blue blouse with lighter blue stripes. Marie's dress may have been modest, but not so that of the wedding reception guests, who were more extravagantly clad. After the civil ceremony at the Town Hall in Sceaux, attended by Marie's father and sister Helena from Warsaw, Bronia and Kazimierz, and the Curie family, Marie and Pierre left on their untraditional honeymoon. They had given each other bicycles purchased from a wedding gift from a cousin. After the reception in the garden of the Curie family house, the newly married couple left to explore Brittany on their shiny new bicycles.

By the 1890s when Marie and Pierre went on a cycling trek for their honeymoon, bicycles had become a fad. The "safety bicycle" with its two wheels of the same size had generally replaced the awkward high-frontwheeled vehicle. When the safety bicycle was first invented, enthusiasts of the high bicycle jeered at their cousins on the safety bikes, claiming that these bikes were ugly and fit only for the lily-livered rider. But soon, the safety bicycle outsold the older form. New groups of people began to take up cycling. Not only was it used for exercise, sport, racing, and touring, but also as a means of transportation for getting to business and social engagements. It appealed to the rich and the not so rich, the long- and short-legged rider, and to both men and women. The high bicycle was indeed dangerous. Reports of cycling calamities appeared in newspapers. The new safety bicycle—with its chain-driven wheels and pneumatic tires—was indeed much more comfortable and considerably safer.

Pierre and Marie had their picture taken on their safety bicycles and both were dressed in recommended cycling costumes. Marie could have cared less about fashionable cycling clothes, but was delighted to find an excuse to toss away her skirts and replace them with comfortable knickers, knee stockings, and low rubber-soled shoes. These bicycles were probably the best purchase that Marie and Pierre could have made. On weekends and on holidays, the young couple took to their bicycles and explored the countryside.

THE EARLY YEARS OF MARRIAGE

Aside from the bicycle trips, about the only recreation Pierre and Marie enjoyed during their first years of marriage was visiting her sister's family in Paris and Pierre's parents in Sceaux. Their first home was a three-room apartment not far from the School of Physics. Marie claimed that its chief charm was its view of a large garden. Since their finances precluded them from having servants, Marie assumed most of the household chores. Most of her time, however, was spent studying. During the first year, she prepared for the teacher's certificate, which would make it possible for her to teach in a girls' secondary school. After several months of preparation she came out first in the examination in 1896. She also took two courses for her own edification. One of these courses was with an inspiring teacher and theoretical physicist, Marcel Brillouin.

During their early married life Pierre continued his research on crystals. He observed that different faces of crystals develop differently. He wanted to explain the reasons for this differential development. Marie wrote that while he obtained interesting results, he never published them. She explained that after he interrupted his investigations to work on radioactivity he never returned to this subject.

While Pierre was preparing his teaching courses for the School of Industrial Physics and Chemistry, Marie assisted him. He first divided his lectures between crystallography and electricity. However, he soon realized that there was not time to treat both areas adequately. Since electricity had a more practical use, he decided to concentrate his lectures on that subject. Marie proudly wrote that his lectures were the most complete and modern to be found in all of Paris. He worked very hard on his lectures, wanting to make sure that they were clear and inclusive. Although he had planned to produce a book from these lectures, his work on radioactivity prevented him from doing so.

Marie was given permission to work at the school with Pierre, although she would have to finance her own proposed research. She found that Pierre's knowledge and experience broadened her own comprehension. Her research project resulted in the completion of her first paper in the fall of 1897.[10] It involved the way in which the magnetic properties of various tempered steels varied with their chemical composition. She was supplied with free samples of steels and had the advice of a leading physicist, Pierre, and a leading chemist, Henri Le Chatelier (1850–1936). Although the paper lacked originality, it gave her the type of experience she would need to pursue her next, highly creative project. In a letter to Józef she wrote that the work on magnetism was part scientific and part industrial. She recognized that it was not a particularly novel paper, but, as she mentioned, it allowed her to work in a laboratory and it was better than giving lessons to students.

By working on the routine project on steel, Marie was exploring a field in which it was improbable that a woman could succeed. Like Marie, other women who loved science often found themselves forced into working in repetitious scientific fields. Although women participated in all aspects of nineteenth-century science, most of them were engaged in data-gathering rather than idea-creation components of science. Significantly, notable exceptions occurred as the century matured. When Curie was working, in the late nineteenth and early twentieth centuries, more women participated in the theoretical sciences than had done so previously. Since most women interested in science lacked university educations, they tended to cluster around fields that did not have specific educational requirements. In the observational sciences, such as botany and astronomy, the expertise of amateurs was appreciated. Women could not only make useful contributions to these subjects but could also remain at home while doing so. Astronomy was one of the few fields that offered women the possibility of jobs outside of the home. But the positions that they were able to secure were those that their male colleagues did not want. The adoption of the technology of cameras and spectroscopes had great implications for women since it required a different labor force. Low-paid positions as "computers" at the Harvard College Observatory and the Royal Greenwich Observatory in England provided women with paid employment. At Harvard, the director, Edward Pickering, hired women because astronomy was moving away from observational astronomy and into the new field of photographic astrophysics. Pickering needed fewer observers (men's work) and many more assistants (women's work) to classify as cheaply as possible the thousands of photographic plates his equipment was generating. And, of course, the women worked for less.

Women were very creative in developing strategies for working in scientific fields. One such strategy, capitalizing on the consideration of certain aspects of science as "women's work," allowed women to work in fields not held in high regard by men. Home economics was one of those areas—a woman interested in chemistry could get a degree in home economics and get a prestigious job in a university or industry because her subject did not interest men.

Another strategy women used to gain a foothold in the sciences was collaboration with a husband or another male relative or mentor. Often women who, on their own, could never get laboratory space, obtain needed equipment, or have their work accepted on its merits were able to accomplish important feats when they had the prestige of a husband, brother, son, or other male investigator, behind them. A woman as a part of a couple could accomplish much original work, although society too often assumed that the creative work was done by the male partner. Some examples of husband/wife creative couples of this period were the British astronomers Annie and Walter Maunder and Margaret and William Huggins; American naturalists Anna Botsford Comstock and John Henry Comstock; French neurologists Cecile Mugnier Vogt and Oskar Vogt; and British physicists Hertha Marks Ayrton and W. E. Ayrton. The most famous collaboration was, of course, that between Marie and Pierre Curie.

Marie Curie faced the same problems that other women scientists had encountered. Her solution, however, was somewhat unique. In order to have time for her studies and research, Marie decided to eliminate all nonessential parts of her life. Although she was a notoriously poor cook, she taught herself to produce passable meals. However, the dishes that she invented needed little preparation or could be left to cook all day while she was at the school. The furniture in their apartment was minimal. They refused to accept the furniture offered to them by Pierre's father, because "every sofa and chair would be one more object to dust in the morning and to furbish up on days of full cleaning."[11] Marie probably realized that a traditional marriage would have drained all of her energy. "Instead the Curies calculatedly pared their family life down to the essentials, thus freeing Marie Curie for a scientific career."[12]

In her biography of Pierre, Marie described their early married years with their common interests "in our laboratory experiments and in the preparation of lectures and examinations." She explained that "during eleven years we were scarcely ever separated, which means that there are very few lines of existing correspondence between us, representing that period." They did, however, take vacations, "walking or bicycling either in the country near Paris, or along the sea, or in the mountains." They never stayed away from Paris for very long, because Pierre found it difficult to be absent for any length of time in a place where he lacked the facilities to work. Nevertheless he, as did she, enjoyed their long walks together, "but his joy in seeing beautiful things never drew his thoughts away from the scientific questions that absorbed him."[13]

As difficult as it was for a woman to be a scientist, it was even more unthinkable that a woman could be both a scientist and a mother. In 1897 when Marie found that she was pregnant with her first child, it would have seemed that her scientific career was over. To make things worse, she was miserable during the early months of the pregnancy. As she wrote to her friend Kazia on March 2, 1897,

> I am going to have a child, and this hope has a cruel way of showing itself. For more than two months I have had continual dizziness, all day long from morning to night. I tire myself out and get steadily weaker, and although I do not look ill, I feel unable to work and am in a very bad state of spirits.[14]

Her pregnancy also made it difficult for her to work on her research project, and she complained that she was vexed at not being able to stand before the apparatus and study the magnetization of steel.

During the same time that Marie was about to give birth to a new life, another life was about to come to an end. Pierre's beloved mother fell ill, and Pierre spent much time with her and away from Marie during the pregnancy. In a letter to her brother, Józef, Marie wrote:

> My husband's mother is still ill, and as it is an incurable disease (cancer of the breast) we are very depressed. I am afraid, above all, that the disease will reach its end at the same time as my pregnancy. If this should happen my poor Pierre will have some very hard weeks to go through.[15]

Pierre finally left his mother to take Marie, who was eight months pregnant, on a bicycling trip to Brest. Both Pierre and Marie seemed to be unaware that such a cycling trip so late in pregnancy was, to say the least, unusual. Although he was normally very considerate, Pierre seemed to expect her to go at the usual pace. They eventually cut the trip short and returned to Paris. Marie gave birth to Irène on September 12, 1897. Pierre and Marie had spent the summer preparing for the 6.6-pound Irène. Marie's fear about her mother-in-law's condition was realized. Two weeks after Irène was born, Pierre's mother died.

Finding adequate childcare has always deterred women scientists from continuing to work after their children are born. Since the Curies always struggled financially, Marie's career might have ended with Irène's birth if it were not for her father-in-law, Eugène. A new widower, Eugène moved in with his son, daughter-in-law, and granddaughter. Eve Curie noted that Dr. Eugène had attached himself "passionately" to the new baby. Marie had first unsuccessfully tried to nurse Irène, but was forced to hire a wet nurse to feed the baby. After a series of nurses and domestic servants she was relieved to turn Irène's care over to her adoring grandfather.

PARIS AT THE TURN OF THE CENTURY

Marie and Pierre were somewhat detached from the vibrant, ostentatious, and often vulgar aspects of turn-of-the-century Paris. Although they did not participate in its excesses, they could not help but be affected by the culture. Science often flourishes in a time of general intellectual excitement. Advances in the arts sometimes precede those in the sciences. This situation applied to France in the late nineteenth century, where Paris was the artistic capital of the world. During the middle to late nineteenth century, French impressionism had its impact on the rest of the world in art as well as music and literature. Among the artists, Edgar Degas (1834–1917), Claude Monet (1840–1926) Auguste Renoir (1841–1919), and Vincent van Gogh (1850–1906) were some of the best known. Claude Debussy's (1862–1918) *Prelude to the Afternoon of a Faun* (*Prélude à l'après-midi d'un faune*) and the poetry of Stéphane Mallarmé (1842–1898) represented impressionism in other fields. In addition to impressionism and postimpressionism, the popular art of Henri Toulouse-Lautrec (1864–1901) memorialized the seedier aspect of Parisian life in his paintings. Toulouse-Lautrec spent much of his time in the Montmartre section of Paris, the center of cabaret entertainment and bohemian life.

As Toulouse-Lautrec was popularizing cabaret life, new technologies invaded Paris and captivated its citizens. Although Marie Curie, the dour Polish woman, and Pierre, her idealist husband, remained aloof from many of the modern trends that showed up everywhere, it was impossible to miss the tallest structure in the world, the Eiffel Tower, which was erected for the International Exhibition of Paris of 1889. The gaslights of the streets of Paris were beginning to be replaced by electric ones. Telephones, moving pictures, and electric streetcars all helped make Paris a modern, exciting city.

The blending of new technology with a surface joviality hid some of the darker aspects of this Paris—a fear and hatred of foreigners, anti-Semitism (a hatred of Jews), and anarchy. The fears festered just beneath the surface hilarity. The French ego had been badly bruised after France was defeated in the Franco-Prussian War (1870–1871). The Prussian chancellor, Otto von Bismarck, was an adept military strategist as well as an adroit politician. France under Napoleon III, on the other hand, seemed both inefficient and inept. When the French did not get certain concessions from Bismarck, they declared war on Prussia and were soundly defeated, forcing Napoleon III to step down. The final rout occurred in Paris on January 28, 1871. Rebels in Paris formed the Commune of Paris and refused to disarm and submit to the French interim government supported by the Prussians. Those who were loyal to the new French government and those who supported the Commune engaged in a bloody battle, which resulted in the vicious suppression of the Commune and eventually the establishment of the Third Republic (1870–1940).

In 1894 widespread French anti-Semitism came to a head. A Jewish captain, Alfred Dreyfus, was accused of spying for Germany. The only evidence against him was a scrap of paper found by a cleaning woman in a wastebasket. Since he was the only Jewish member of the general staff, he was immediately suspect. Dreyfus was convicted on forged evidence and sentenced to life imprisonment on Devil's Island off the coast of South America. The rabidly nationalist press continued to condemn Dreyfus as a traitor. Even after the real culprit was discovered, a Major Esterhazy, the crowds continued to malign Dreyfus, and it took the fall of the government to free him.

Alongside the chaos accompanying a French society that was unhappy with its government, disliked and feared foreigners, and was notably anti-Semitic, an anarchist movement blossomed. Anarchism is a political theory that finds all forms of governmental authority both unnecessary and undesirable. Ideally, it results in a society based on voluntary cooperation and the free association of individuals and groups. Some French intellectuals supported the French version of this international movement. Unfortunately, the ideals of the theory became subverted and sometimes resulted in violence. In a two-year period between 1892 and 1894, eleven anarchist bombs exploded in Paris.

Science was also criticized by those who were dismayed at the turn French institutions had taken. Many people thought that science, with its emphasis on reason and its apparent worship of positivism, seemed to support the antichurch leaders of the Third Republic. Scientists tended to make unrealistic claims for science and technology, yet some of the most exciting scientific breakthroughs were born in this period. Marie and Pierre Curie were scientists who thought that salvation lay in science and reason.

NOTES

1. Susan Quinn, *Marie Curie: A Life* (New York: Simon and Schuster, 1995), p. 101.

2. Marie Curie, *Pierre Curie*, trans. Charlotte and Vernon Kellogg (New York: Macmillan, 1923), p. 42.

3. Ibid., p. 51.

4. Eve Curie, *Madame Curie: A Biography* (Garden City, NY: Doubleday, Doran & Co., 1938), p. 126.

5. Marie Curie, *Pierre Curie,* p. 77.

6. Robert Reid, *Marie Curie* (London: Collins, 1974), p. 65.

7. Quinn, *Marie Curie: A Life*, p. 117.

8. Marie Curie, *Pierre Curie,* p. 80.

9. Eve Curie, *Madame Curie*, p. 137.

10. Marie Curie, "Propriétés des aciers trempés," *Bulletin de la Société d'encouragement pour l'industrie nationale* (1898), in *Oeuvres de Marie Slodowska Curie,* ed. Irène Joliot-Curie (Warsaw: Panstwowe Wydawnictwo Naukowe, 1954), pp. 3–42.

11. Eve Curie, *Madame Curie*, p. 143.

12. Helena M. Pycior, "Marie Curie's "Anti-Natural Path," in *Uneasy Careers and Intimate Lives: Women in Science, 1789–1979,* ed. Pnina G. Abir-Am and Dorinda Outram (New Brunswick, NJ: Rutgers University Press, 1987), p. 199.

13. Marie Curie, *Pierre Curie*, p. 82.

14. Eve Curie, *Madame Curie*, p. 147.

15. Ibid., p. 148.

Chapter 5

THE DISCOVERY OF RADIUM: A SCIENTIFIC BREAKTHROUGH

Believing as she did in the importance of science, Marie Curie was determined to obtain a doctoral degree—a requirement if her research was to be respected by her peers. At this time no woman in Europe had completed this degree. An unmarried German woman, Elsa Neumann (1872–1902), was writing a thesis in electrochemistry and would eventually finish it, but it seemed impossible for Curie to complete the arduous research necessary for the degree. The challenges that she faced as wife, mother, as well as scientist appeared insurmountable. Even more daunting was the disapproval of her colleagues who were convinced that a married woman, and especially a woman with children, could never earn a doctorate. Nevertheless as she prepared her monograph on steels for publication, she began to search for a suitable topic for her doctoral thesis. Familiar with the scientific literature, she was aware of Wilhelm Röntgen's (1845–1923) discovery of x-rays on November 8, 1895.

Röntgen realized that the name electrode referred to a conducting material used to make electrical contact with part of a circuit and could be charged either positively or negatively. A positively charged electrode is called the anode and the negatively charged electrode (the source of what we now call electrons) is known as the cathode. While head of the physics department at the University of Würzburg, Röntgen investigated the properties of cathode rays, or negatively charged particles (electrons) emitted by a high-vacuum discharge tube, which had been perfected in the 1850s. When activated by a high-voltage current, the electrons would race from the cathode to the anode. Sometimes the electrons were invisible and at other times they appeared as blue streaks. When the rays touched the glass wall of the tube, they created a green or blue luminescence. Röntgen became interested in the luminescence (glow) that the cathode rays produced. Röntgen tried to reproduce the work of a German researcher, Phillip Lenard (1862–1947), who had observed the behavior of cathode rays when they escaped from the vacuum tube. He found that they would illuminate a substance some distance away from it when coated with a phosphorescent material (a substance that emits light without appreciable heat). This phenomenon fascinated Röntgen, who tried to repeat and modify the experiment. His method of studying the phenomenon was to wrap the cathode ray tube in black cardboard to exclude all light, darken the room so that he could see a faint light, activate the tube with a high-voltage current, and observe the gleam. When he activated the tube and observed a flash of light, he unexpectedly found that the light did not come from the tube. He observed that a sheet of paper coated with the compound barium platinocyanide glowed (luminesced or phosphoresced) even when the tube was blocked off by the black cardboard and could not possibly have reached the barium platinocyanide.

When the coated part of the phosphorescent screen was turned away from the discharge tube, it still phosphoresced. Röntgen postulated that the resultant radiation could not have been caused by the cathode rays because these rays could not penetrate the cardboard. He turned off the tube and the coated paper darkened. He turned it on again and it glowed. Uncertain as to whether to believe his own eyes, he carried the coated paper into the next room, closed the door, and pulled down the blinds. As long as the tube was in operation, the paper continued to glow. These rays apparently had the ability to penetrate substances and even pass through walls. He established that these rays passed unchanged through cardboard and thin plates of metal and were not deflected by electric or magnetic fields, as were cathode rays. Although he clearly had observed a new type of radiation, he was unable to establish its nature and coined the term x-rays (unknown rays) for the new rays. These rays actually came from the glass walls of the tube when struck by cathode rays. In subsequent experiments he immobilized his wife's hand over a photographic plate in the path of the rays. After he developed the plate he saw an image of her hand, which showed the shadows thrown by the bones of her left hand and that of a ring as a dark blob on her fourth finger. Although it is probable that x-rays had been produced by others before Röntgen, he first realized their existence and first investigated their properties.

Röntgen's breakthrough excited laypeople and scientists alike and netted its discoverer the first Nobel Prize in Physics in 1901. The most spectacular property of these rays was their ability to penetrate flesh but not bone. The medical and entertainment possibilities seemed endless. At a demonstration of x-rays at the Royal Society on May 6, 1896, the scientist Lord Kelvin's hand was x-rayed illuminating the underlying bones. The idea that a process had been found that enabled one to see through solid objects seemed marvelous indeed. Röntgen became a reluctant hero, for it was no longer possible for him to do his research in solitude and quiet. Exciting medical potentials for the new ray were apparent, but the possibility of harmful results was unsuspected. Frivolous uses were common. One such device, a fluoroscope, was used to assure that children's shoes fit properly. Most shoe stores had one, and children were amused by looking at the bones of their feet. By stepping on a platform with new shoes on, a child and his/her parents could look through a lens and see these bones. No one had any idea that danger might be involved. In the meantime, several generations of children were entertained by looking through a viewer at their feet encased in new shoes and being amazed at seeing the underlying bones.

Although shortly after the original discovery was made several scientists postulated that x-rays were electromagnetic rays similar to visible light but with shorter wavelengths, the actual nature of these rays was not firmly established until eighteen years later. The search for proof provided research opportunities for numerous scientists and technicians. These new rays caught the imagination of the public, and many popular articles appeared. Scientists, too, jumped on the bandwagon. Many people joined the search for another heretofore undiscovered new form of radiation. Two French investigators, Gustave LeBon (1841–1931) and René Blondlot (1849–1930), claimed to have made such a discovery. LeBon had named his new radiation "black light," and his discovery was not given much credence by other investigators. Blondlot, on the other hand, claimed that he had produced his "N rays" (named after his native city of Nancy) by placing a hot wire inside an iron tube. The rays were then detected by a calcium sulfide thread that glowed slightly in the dark when the rays were refracted through a prism of aluminum with sides angled at sixty degrees. Blondlot claimed that a narrow stream of N rays were refracted through the prism and produced a spectrum. The N rays were invisible except when they encountered the thread. Confirmation quickly followed, as scientists from laboratories all over the world claimed to have generated N rays.

The British journal *Nature* was suspicious of Blondlot's claims because German and English laboratories had not been able to replicate his results. The journal sent the American physicist Robert W. Wood of Johns Hopkins University to investigate. Unbeknownst to Blondlot or his assistant, Wood removed the prism from the detection device. When Blondlot's assistant conducted the next experiment he found N rays, which, of course, he should not have. Wood tried to sneak the prism back in place, but the assistant saw him and thought he was removing it. The next time he tried the experiment the assistant averred that he could not see any N rays, but of course he should have if the experiment was valid. The N ray episode does not mean that Blondlot was attempting to deceive, nor should blame be directed entirely toward the assistant. These points are important to remember today, for it still happens that when a scientist badly wants something to be true, he or she may ignore evidence to the contrary. An example is the cold-fusion fiasco of 1989. Fusion is the process that takes place in the sun's core where at extraordinary high temperatures hydrogen atoms are compressed to form helium and a massive amount of energy. This is the same kind of thermonuclear explosion that a hydrogen bomb releases. If a controlled form of fusion could be discovered, an unlimited, cheap, pollution-free form of energy would be available. Two experimenters at the University of Utah, Martin Fleishmann and Stanley Pons, thought that they had achieved this fusion in the laboratory at room temperature. The two investigators rushed into print and several other laboratories reported the same results. The consensus among scientists was that the two investigators had not deliberately perpetrated a hoax, but that they saw what they wanted to see.

Although the N ray claims of LeBon and Blondlot were spurious, the discovery of another new type of radiation was not. After Röntgen's

1896 paper was published, Henri Poincaré (1854–1912) attempted to explain it in a report to the French Academy of Sciences. Much like Röntgen, Poincaré was fascinated by the process known as phosphorescence—the glow caused by light on certain substances. Even after he removed the light from these substances, the glow continued. Poincaré noted that x-rays caused phosphorescence both on the wall of the vacuum tube and on a screen outside the tube, which was coated with a phosphorescent substance. Another scientist, Alexandre-Edmond Becquerel (1820–1891), had invented an instrument to identify new substances with phosphorescent qualities—even when they only phosphoresce for a very short time. Although Alexandre Becquerel was dead before the meeting when Röntgen reported on his new x-rays, his son, Antoine Henri (1852–1908), listened to Röntgen's and Poincaré's results and became fascinated with phosphorescence, his father's interest. Although Henri had a doctorate from the Sorbonne and was a member of the academy, he was not active in research until he heard Poincaré's report on x-rays.

Becquerel and three other scientists hypothesized that the phosphorescent substance itself could produce x-rays and that the cathode ray tube was unnecessary. The three other scientists were convinced of the correctness of their hypothesis and found evidence to corroborate it. Henri Becquerel, on the other hand, did not find x-rays when he experimented on phosphorescent substances. After negative results, he tried a different phosphorescent substance—a sample of uranium salts. These salts immediately produced radiation. Becquerel prepared a written report to the French Academy and explained his methodology. He reported that he had taken a photographic plate and wrapped it in two sheets of thick black paper to protect the plate from the sunlight. Then he placed a plate of phosphorescent substance above the paper and exposed the entire package to the sun for several hours. He observed that when the photographic plate was developed the silhouette of the phosphorescent substance appeared in black on the negative. He then placed a coin between the phosphorescent material and the paper, and exposed it to the sun and found that its image appeared on the negative. He concluded that the phosphorescent substance that he used emitted rays that could penetrate paper impervious to light.

Becquerel assumed that it was the sun that allowed the material to phosphoresce and to penetrate the photographic plate. But being a careful scientist, Becquerel went to his laboratory to prepare another experiment to corroborate his first results. For this trial, he placed a thin copper cross between the black paper covering the plate and the uranium salts. He, of course, postulated that when the package was exposed to the sun, the pattern of a cross would appear on the plate. However, bad weather in Paris during February delayed the experiment, because the sun steadfastly refused to shine. Becquerel placed the entire setup in a dark cabinet until the weather changed for the better. Though the sun still declined to appear, Becquerel became impatient and developed the plate. Since the sun was supposedly the agent that would cause the plate to darken, Becquerel was amazed to find that the plate was not blank, as he had assumed that it would be. Instead, the plate had darkened as if it had been exposed to sunlight, and the image of the cross stood out in white against the black background. He was forced to conclude that sunlight was not necessary for the impression to appear on the photographic plate. He concluded that it was the uranium in the mixture that caused the reaction. Continuing to explore the situation in his next four papers, Becquerel was convinced that it was the uranium that caused the image to be produced on the photographic plates. However, he never gave up his previous conception that phosphorescence was involved in some way in the phenomenon. He assumed that an energy (which he called a form of phosphorescence) was stored in the uranium and concluded that the emission produced by the uranium was the first example of a type of invisible phosphorescence. Although he had actually discovered radioactivity, Becquerel neither named it nor explained its source. Physicist Jean Perrin (1870–1942) concluded that Becquerel was a prisoner of the previous hypothesis of phosphorescence, although he was able to move further from it than the other three scientists.

Giving up a well-entrenched idea is one of the most difficult things for a scientist to do. In order to keep the established hypothesis he or she will sometimes ignore contradictory results or will add new postulates in order to save the theory. Thus in Becquerel's case, he was convinced that phosphorescence was involved even when the evidence he went out of his way to collect seemed to indicate that something different was happening.

Becquerel was not the only one experimenting along these lines.

While Becquerel was working in France, Silvanus P. Thompson (1851–1916) in London put a small quantity of uranium nitrate over an aluminum-covered photographic plate and observed its effect. After putting the plate on the window sill he developed the plate and discovered it had darkened at the place where the uranium salt had been. Surprised that uranium could affect his plate through the aluminum shield, he wrote to George Stokes (1819–1903), the president of the Royal Society, to tell him about his results, which he christened hyperphosphorescence. Stokes was enthusiastic and urged him to publish immediately. But unfortunately for Thompson, Becquerel published first and received the credit. As so often happens in science, it seems that the time was ripe for an idea. Discoveries in science are usually very complex occurrences based on ideas, instruments, or techniques developed in specific social and cultural situations. When Sir Isaac Newton (1642–1727) postulated his gravitational hypothesis, Robert Hooke (1635–1703) and others were working on the problem and had worked along very similar lines to Newton, although not as completely. When Charles Darwin (1809–1882) produced his hypothesis of evolution through natural selection, Alfred Russel Wallace (1823–1913) had arrived at a very similar hypothesis. Although it is possible to find information on these "also rans," they seldom get the credit that they deserve. Thus by a quirk of luck, it is Becquerel, not Thompson, who is remembered and his experiment cited as producing the seminal work in radioactivity. Recognition by contemporaries reflects the importance of a discovery on subsequent researchers. For example, although Leonardo da Vinci's (1452–1519) science is recognized today it had little impact on Renaissance science because he kept his discoveries secret. Thompson did not deliberately keep his work secret but because Becquerel was better known, he got the credit. However, unlike Röntgen's x-rays, Becquerel's experiment did not get immediate attention. One man in particular, though, William Thomson, Lord Kelvin (1824–1907), furthered Becquerel's experiment by asking whether uranium rays electrified the air, as did x-rays. Using his electrometer (an instrument for the measurement of electric currents), Kelvin confirmed that they did.

Becquerel's new kind of ray intrigued Marie Curie, as she was searching for a subject for her doctoral thesis. She found it interesting that Becquerel's rays had not appealed either to popular or to scientific sensibilities. She wrote in her autobiography that she and her husband were excited by the new phenomenon and that she resolved to "undertake the special study of it."[1] She wanted to understand the nature of Becquerel's radiation. However, in order to explore these rays she had to have a suitable laboratory available. After Pierre approached the director of his school, Marie was given the free use of a glassed-in room on the ground floor that was more like a storeroom than a laboratory. The room was damp and humid in the summer and bitterly cold in the winter. It was totally unsuitable for scientific instruments or for Marie's health.

Marie first planned to detect a way of determining whether other substances besides uranium caused the air to conduct electricity. She used two of Pierre's inventions, the electrometer and the piezoelectric quartz balance, to test the substances. Together, she and Pierre developed a technique. She put the powdered uranium on one metal plate and opposed it with a second plate. The plate with the uranium was charged, and she used the electrometer to determine whether an electric current passed through the air between the plates. After she had perfected the technique she began to test other elements to see if any had properties similar to uranium. After testing dozens of materials, she found that thorium and its compounds caused the air to conduct electricity and produced rays similar to those from uranium.

Her doctoral project began to change from merely a descriptive one in which she measured the electric current emitted by different materials to a theoretical one where she speculated on a cause for the radiation. In a very understated way, she reported her preliminary results on uranium as follows:

> My determinations showed that the emission of the rays is an atomic property of the uranium, whatever the physical or chemical conditions of the salt were. Any substance containing uranium is as much more active in emitting rays, as it contains more of this element.[2]

In other words, she realized that the activity of the uranium compounds depended solely on the amount of uranium present. It did not matter if the uranium salt was dry or wet, lumpy or powdery, or what other elements were present in the salt. If she did nothing else, this discovery would place her in the ranks of first-rate scientists. She had demonstrated that the radiation was not caused by an interaction between molecules—it was not an ordinary chemical reaction where light or heat was given off as a product of the reaction. It issued from the atom itself. Radiation was an atomic property, proportional to the amount of the radioactive substance being measured.

At this time she did not speculate on the significance of this interpretation, and continued to measure the conductivity of the air using two other minerals containing uranium, chalcolite, and the uranium ore, pitchblende. She determined that pitchblende was four times as active as uranium, and chalcolite two times as active. She concluded that these two substances must contain other substances that were much more active than uranium. In her autobiography she made it evident that this idea was hers and not Pierre's. When writing about it she uses the first person singular pronoun, "I."

> There must be, I thought, some unknown substance, very active, in these minerals. My husband agreed with me and I urged that we search at once for this hypothetical substance, thinking that in beginning this work we were to enter the path of a new science which we should follow for all our future.[3]

In addition to analyzing the pitchblende by known chemical methods, the couple used Pierre's delicate electrical apparatus to examine different portions for evidence of radioactivity. Marie was positive that the strong radiation that they observed came from a new chemical element. She confided her conviction to Bronia. She also told her sister that the physicists with whom they had spoken were convinced that they had made an error in experimentation. Marie realized that only by actually isolating the element would she ever persuade the skeptics.

The general method that they followed was to grind up the pitchblende and then dissolve it in acid. Afterward, they broke it down into different components using standard chemical techniques. They then measured the radioactivity of the products. The next step was to use spectroscopic analysis to attempt to find the highly active unknown element. Robert Bunsen (1811–1899) and G. R. Kirchhoff (1824–1887) first developed a method for using spectroscopy to identify unknown elements in the 1860s, and since that time eight previously undescribed elements were discovered. Their friend Gustave Bémont, who had the laboratory expertise that they lacked, heated a new sample of pitchblende in a glass tube and distilled a small quantity of the material on the glass. This material was far more active than pitchblende. However, when the Curies tested their sample, it was not sufficiently pure to show the characteristic spectral lines of a new element. They did not scrap their new element hypothesis, but assumed that they had to produce a purer form from the pitchblende in order to demonstrate its presence.

Marie tried a tedious technique called fractional crystallization to separate out different substances from a solution of pitchblende. This procedure depends on the fact that different substances in the same solution form crystals at different temperatures. Those with lower atomic weights crystallize first. Marie first boiled the pitchblende solution, then gradually cooled it, and finally tested the crystals that were formed for radioactivity with the Curie electrometer. She discarded the crystals that were formed first, which were not radioactive or only slightly so. She repeated this technique over and over again on the solution, retaining the more radioactive fraction and discarding the less active crystals. With each fractional crystallization, the crystals became increasingly more radioactive.

The element bismuth in the pitchblende stubbornly refused to be separated from what Marie presumed was the new element. Marie was just as obstinate as the clinging bismuth, which she attempted to separate. On June 6, 1898, she took a solution of bismuth nitrate that she was convinced contained her new element and added hydrogen sulphide to it. She collected the solid (bismuth sulphide) that was precipitated and measured its activity and found that it was "*150 times more active than uranium.*"[4] Not happy with the impure product, bismuth sulphide, that they had collected, Pierre heated a small sample of this substance in a glass tube. The bismuth sulphide remained in the hottest parts of the tube, while a black powder was deposited on the glass at slightly cooler temperatures. He and Marie continued working and eventually came up with a product that was 330 times more active than uranium. Although they still did not obtain the pure form, they were convinced that they had a new element. Marie wrote, "In July, 1898, we announced the existence of this new substance to which I gave the name of polonium, in memory of my native country."[5]

In the paper that they produced in 1898, the term "radioactive" was first used to describe the behavior of uranium-like materials. Because of the connection with her native country, Marie felt that the discovery of polonium would be her most significant contribution. However, the entries in their laboratory notebook in November make it very clear that they had previously made another discovery that was to be even more significant. They found that the remaining liquid when they had gotten rid of both bismuth and polonium was still extremely radioactive. The remaining impurity in the liquid was the element barium, which was known not to be radioactive. Concluding that they had discovered not only one but two radioactive substances in the pitchblende they named their second new element radium. Radium was far more radioactive than polonium and 900 times more radioactive than uranium. It was to be radium, not polonium, that gave Curie her instant and lasting fame.

It is one thing to postulate new elements and still another to actually isolate them. In order to convince chemists of the correctness of her assumptions, Marie felt it necessary to produce a pure form of the new elements. Pierre, whose background was in physics, was more content to rely on his mind rather than his senses to tell him that a new element was the most reasonable explanation for the radiation. Thus, it was Marie who felt it necessary to go through the tedious procedures in order to have physical evidence of their hypothesis. Before they would believe that she actually had discovered a new element, her fellow chemists wanted her to ascribe an atomic weight to polonium and radium. She could only find these weights if she could isolate the pure elements. As they worked toward finding the unknown substances, the Curies did not know any of its chemical properties. Since they only knew that it emitted rays, the rays represented their starting point.

In order to isolate the new elements, Marie needed a large quantity of a source material. Pitchblende was a heavy gooey black compound that was mined on the German-Czech border in the Joachimsthal region. This region first became famous for its silver mines. The discovery of silver led to the minting of about two million large silver coins called Joachimsthaler. The name was later shortened to thaler, which is the source for our word dollar. In the late eighteenth century a chemist, Martin Heinrich Klaproth (1743–1817), extracted a gray metallic element from one of these mines. He named this element uranium after the planet Uranus that the astronomer William Herschel (1738–1822) had recently discovered.

Uranium was important economically because it produced a superb agent for coloring ceramic glazes. It gave a velvety finish to porcelain. It also was sometimes used to toughen steel.

It was not the practical use of uranium that interested the Curies. Marie, with Pierre's blessing, set out to hunt for the active substance in the uranium ore, pitchblende. Because of its practical utility, pitchblende was very expensive. The Curies realized that they would have to treat huge quantities of the material in order to isolate their new metals. Overcoming obstacles was nothing new to Marie and Pierre. They assumed that after the valuable uranium was extracted from the ore, traces of polonium and radium would remain in the residue. Useless for most purposes, this residue was very cheap. Still, they realized that it would be futile to ask the University of Paris or the French government for a grant to buy it, for they were notoriously stingy. After getting permission from the directors of the mine of St. Joachimsthal, the Curies went to their meager savings and retrieved enough money to buy the crude material and pay for its transportation to Paris.

Their next problem was to find a place where they could store huge quantities of the radioactive substance. When it arrived in 1899, they had it dumped in the yard of the School of Physics. The storeroom that they had previously used for their research clearly was inadequate for the deposited sacks of pitchblende. The new director of the School of Physics was not nearly as tolerant and helpful as the previous one had been, and if it had not been for Marie's quiet insistence he probably would not have complied with their request for a different space. An abandoned shed with a dirt floor, formerly used as a medical school dissecting room, was put at their disposal. The sacks remained in the yard and the shed served for the analytical work. Marie observed that it

> surpassed the most pessimistic expectations of discomfort. In summer, because of its skylights, it was as stifling as a hothouse. In winter one did not know whether to wish for rain or frost; if it rained, the water fell drop by drop, with a soft, nerve-racking noise, on the ground or on the worktables, in places which the physicists had to mark in order to avoid putting apparatus there. If it froze, one froze.[6]

Since to Pierre Curie it seemed superfluous to engage in the enormous physical struggle to demonstrate what they already knew, much of the exhausting labor was left to Marie. When the physicist Georges Urbain (1872–1938) returned from a visit he reported that he "saw Madame Curie work like a man at the difficult treatments of great quantities of pitchblende." She moved the heavy containers, transferred the contents from one vat to another, and, "using an iron bar almost as big as herself," spent "the whole of a working day stirring the heating and fuming liquids."[7] They worked under these circumstances from 1898 to 1902. Even under these abominable conditions, Marie noted that it was in this "miserable old shed" that she experienced the "best and happiest years of our life."[8]

By 1901, the Curies had divided their research into two parts, the isolation of radium and the study of the rays associated with radioactivity. Marie embraced the methods of the chemist, and Pierre, tired of the endless extraction process, preferred the methods of the physicist. In 1898, Pierre, Marie, and Bémont noted in *Comptes Rendus,* a journal of the French Academy of Sciences,

> Two of us have shown that by purely chemical processes one can extract from pitchblende a strongly radioactive substance. This substance is closely related to Bismuth in its analytical properties. We have stated the opinion that pitchblende may possibly contain a new element for which we have proposed the name *polonium.*[9]

In an earlier article, the Curies had announced the existence of polonium, but in this later article, besides reminding their readers of polonium, the Curies also laid the foundation for their future assertion that pitchblende contained another extremely radioactive element, radium. They described the properties of radium and explained why it could not be a previously known element. They concluded in this paper "the various reasons which we have just enumerated lead us to believe that the new radioactive substance contains a new element to which we propose to give the name *radium.*"[10]

Persistent Marie continued her efforts to purify radium. Finally in 1902 after treating over a ton of pitchblende residues, Marie produced one-tenth of a gram of almost pure radium. She prepared it from a radium salt (radium chloride) and determined that its atomic weight was 225.93. It pleased both Marie and Pierre to observe that their new element was radiantly beautiful, exuding a blue luminosity that was spectacular. After announcing the result in her own name, Marie began to write her doctoral thesis, "Researches on Radioactive Substances." She defended her thesis on June 25, 1903. In a thesis defense it was customary to have outside people attend in order to support the candidate. More than the usual number of outsiders attended Marie's thesis defense. In the crowded examination hall, curiosity seekers as well as family, friends, and colleagues were present. After the examination she was awarded the degree of Doctor of Physical Science at the University of Paris, with the added accolade of *très honorable.* Her thesis was published in the same year.

Although the isolation of radium scientifically took second place to the early conclusion that radioactivity was atomic in nature, the isolation of the glowing blue radium was the stuff of legends and led others to fall under its charm. In 1900, the Curies collected all of the research on radioactivity that they could find and published it in a long paper. Although they described the properties of these rays, the source of this curious energy remained unknown. They wrote that the spontaneity of the radiation was a subject of great astonishment. It seemed to violate one of the most sacred laws of physics, the first law of thermodynamics. This law states that energy cannot be created or destroyed although it can be converted from one form to another. Radium did not seem to undergo any change but just emitted energy. In this paper they asked questions about the source of energy coming from the rays. Does it originate within the radioactive bodies, or is it imposed from without? Every conclusion that they came up with seemed to violate one of the most important assumptions of nineteenth-century physics. Up until the discovery of radioactivity, physicists could explain all phenomena in nature by gravitational attraction and electromagnetic force. New forces contained within the nucleus of the atom would later be called upon to account for radioactivity, but in 1900 the behavior of radioactive substances was an enigma.[11]

Although most people assume, as did physicist Ernest Rutherford, that the Curies' collaboration consisted of Marie doing the chemistry and Pierre the physics, another scholar sees their collaboration in a different way. Rutherford wrote:

> While at this stage, M. and Mme. Curie did all their scientific work together, it is natural to assume that Mme. Curie, as the chemist of the combination was mainly responsible for the chemical work involved. She alone was responsible for the large-scale chemical work required to separate radium from radioactive residues in sufficient quantity to purify it and obtain its atomic weight.[12]

But it can be argued that their successful collaboration was much more than Marie serving as the chemist and Pierre the physicist, and that their "success as a scientific couple included, but was not limited to, the partners' different commitments to chemistry and physics."[13]

HOME LIFE DURING THE RADIUM YEARS

During the summer of 1898 after the discovery of polonium, Marie, Pierre, and baby Irène went on vacation in Auroux in the Auvergne, a very mountainous region of France. Even though they hiked, swam, and played, Marie and Pierre never left their "'new metals,' polonium and 'the other'—the one that remained to be found—'behind.'" Although they enjoyed leaving the sultry summer heat of Paris, they were pleased to return to their research in September.[14] In the fall of that year, Bronia and Kazimierz left Paris for Poland, leaving Marie without a direct family connection to her beloved Poland. She wrote Bronia, "it seems to me that Paris no longer exists, aside from our lodging and the school where we work." She asked advice about how often the green plant they left behind should be watered. She reported on the family, writing,

> We are well, in spite of the bad weather, the rain and the mud. Irène is getting to be a big girl. She is very difficult about her food and aside from milk tapioca she will eat hardly anything regularly, not even eggs. Write me what would be a suitable menu for persons of her age.[15]

Marie treated her household activities in much the same way that she did her scientific experiments. For instance, she wrote annotations in the margins of a cookbook about making gooseberry jelly.

> I took eight pounds of fruit and the same weight in crystallized sugar. After an ebullition [boiling] of ten minutes, I passed the mixture through a rather fine sieve. I obtained fourteen pots of very good jelly, not transparent, which "took" perfectly.[16]

She recorded every detail of Irène's life much as she recorded meticulously every detail of her laboratory work. In a notebook she wrote that Irène "can walk very well on all fours." She also carefully noted her weight gain day by day, her diet, and the appearance of each tooth. During their summer vacation trip to Auroux, she recorded that Irène "plays with the cat and chases him with war cries. She is not afraid of strangers any more. She sings a good deal. She gets up on the table when she is in her chair."[17]

Both Marie and Pierre were convinced that having a baby need not cause Marie to give up her scientific work. She wrote that "such a renunciation would have been very painful to me, and my husband would not even think of it; . . . Neither of us would contemplate abandoning what was so precious to both." Marie was lucky to have a built-in baby sitter in Dr. Eugène Curie. While Marie was in the laboratory, Irène was in the care of her grandfather, who "loved her tenderly and whose own life was made brighter by her."[18] His help was especially appreciated because Marie was never a skilled homemaker, and could see little virtue in such humble tasks. The strong support that she had from a close family made an ordinarily impossible situation much more tolerable. "Things were particularly difficult only in cases of more exceptional events, such as a child's illness, when sleepless nights interrupted the normal course of life." Marie noted, however, that they had little time for ordinary social relationships. Their friends were scientists like themselves. A social evening would consist of talking with these friends in their home or in the garden. Marie did find time to sew for Irène. She never bought ready-made clothes for her, for she thought them too elaborate and impractical.

Pierre should have been in a good position to finally attain an academic chair. His work on crystallography, piezoelectricity, symmetry, and magnetism were admired and well known. His joint work with Marie resulting in the discovery of radium and polonium was widely accepted. Thus, when the chair of physical chemistry became available at the Sorbonne in 1898, Pierre asked for it and had good reason to think that he would be appointed. Again, he was disappointed and had to settle for the post of assistant professor *(repetiteur)* at the Ècole Polytechnique to supplement their income. If he would have been appointed, the results would have been prestige, a better laboratory, and a modest increase in salary. Pierre, however, was not surprised when he did not get the position of chair. Since he was not a graduate of either the Normal School or the Polytechnic, he suspected that he would be passed over. For without the support that the large schools provide their graduates, nongraduates were often ignored in spite of their accomplishments. In this case a younger colleague, Jean Perrin, who had the advantage of a prestigious education got the appointment.

The Curies' disappointment almost resulted in their leaving Paris. The University of Geneva offered Pierre a chair with a high salary and a laboratory to be designed to his own specifications. The university also promised an official position to Marie. They seriously considered the offer and went to Geneva to evaluate the new opportunity. In her biography of Pierre, Marie noted that it was a "grave decision for us to make." However, although Pierre was tempted by the offer, he finally decided to remain in Paris because he "feared the interruption of our investigations which such a change would involve."[19] It was a fortuitous opportunity that another position opened at the Sorbonne, teaching physics to medical students. With the encouragement of his colleagues, in particular the influential Henri Poincaré who wanted to keep him in France, Pierre was hired. Although this made him a member of the faculty of the Sorbonne, the course served medical students and was peripheral to the prestigious Faculté des Sciences where Pierre wanted to be. Although their income increased, the working conditions were not good. Pierre had a double teaching load and was often fatigued with teaching so many students. One of the worst problems with the new position was that it did not include a laboratory. All that he had at his disposal was a little office and a workroom. Since he was determined to continue with his research, he had to go back and forth on his bicycle between his new office, the makeshift laboratory, and the École de Physique and Chemie. When another chair opened up in mineralogy, for which Pierre was qualified, he was refused a second time. Pierre was simply a poor politician. He refused to flatter those who made personnel decisions and tended to blame his failures on his lack of a prestigious school background.

At the same time that Pierre accepted the physics teaching position at the Sorbonne, Marie also got a new paid position. She was the first woman to be named to the faculty of the Normal School (École Normale Superieure) for Girls at Sèvres. This school was the elite preparatory school for teachers in France. Marie's first year of teaching at the school was difficult, for students made fun of her Polish accent and her somewhat awkward sentence structure. More important, however, it was her dry lecture presentations that acted as a put off for her students. Between her first and second years of teaching she seems to have undergone an epiphany. She used experiments instead of lectures. The students then would discuss the meaning of the experiments. She went from being a pariah to the most popular teacher at the school.

During these years their social life was modest but did exist. They occasionally visited Marguerite and Émile Borel at their frequent evening parties. For a special occasion, Marie would dress up for a night at the theater. When they had a visitor from another country they would go out to lunch and dinner and show him or her the sights of Paris, including a trip up the Eiffel Tower. These occasions were rare, however, and most of their social life consisted of conversations with scientists and close friends who had children about the same age as Irène.

Pierre's friends worked to prepare him for a desirable professorship. They convinced him to present himself as a candidate for membership in the prestigious Academy of Science (Académie des Sciences) where they assured him that his election was certain. Since the physicist members of the academy promised to support him, he reluctantly agreed to apply for membership. However, as in the case of the professorships he was to be disappointed. The process of becoming accepted was demoralizing for the shy and diffident Pierre. It was customary for the candidate to visit the current academy members. He found the entire process humiliating. A journalist later wrote of Pierre's visits to the academicians as follows:

> To climb stairs, ring, have himself announced, say why he had come—all this filled the candidate with shame in spite of himself; but what was worse, he had to set forth his honors, state the good opinion he had of himself, boast of his science and his work—which seemed to him beyond human power. Consequently he eulogized his opponent sincerely and at length, saying that M. Amagat was much better qualified than he, Curie, to enter the Institute.[20]

When the results of the election were received, M. Émile Amagat received thirty-two votes, Pierre twenty, and a third candidate six. Pierre told his friend that he was not disappointed—that the only thing that concerned him was the time spent away from his research making the visits.

Pierre's new dean, Paul Appell (1855–1930), mounted an attempt to get Pierre recognized. In response to a request from the ministry asking the dean to propose names for the Legion of Honor, he wrote Pierre begging him to allow his name to be submitted. In an attempt to convince Pierre to accept the award, Appell wrote to Marie, implying that acceptance would bring a bigger laboratory and the equipment that he so needed. Pierre found it ridiculous to have his laboratory needs contingent on having the Legion of Honor's little enameled cross hung on the end of a red silk ribbon. Therefore, he replied to the dean to "please be so kind as to thank the Minister and to inform him that I do not feel the slightest need of being decorated, but that I am in the greatest need of a laboratory."[21]

As early as 1897, the Curies had begun to have health issues. They, as well as their friends, blamed the problems on overwork and refusal to eat and rest properly. A colleague and friend, Georges Sagnac (1869–1926), complained that they ate very little. He asked if even "a robust constitution wouldn't suffer from such insufficient nourishment?" His suggested cure was to have regular meals and avoid "discussion of distressing or dispiritying [*sic*] events. You must not read or talk physics while you are eating."[22] Since there was a history of tuberculosis in her family, Marie was diagnosed with a suspected tubercular lesion of the lung. However, this problem did not develop beyond the initial symptoms. Although they never acknowledged it, the timing suggests that the decline in their health was related to the new rays that they were studying. Being Marie, she kept an accurate account in her notebook of all of her symptoms.

Sagnac's good advice would have been of little help if it indeed was exposure to heavy doses of radiation that caused their sickness. Although it was understood that radium could cause local burns, the more serious and widespread systemic effects were unknown. Pierre and Henri Becquerel published a paper in 1901 in which they described burns on their skin caused by contact with radioactive material. Becquerel was accidentally burned while carrying a glass tube of radium in the pocket of his waistcoat. Pierre duplicated an experiment by two Germans who were the first to report burns from radioactive materials in print. After placing thickly wrapped radioactive barium on his arm for ten hours, he observed that the skin was red. The redness increased for several days and on the forty-second day the skin began to heal around the edges of the wound. On the fifty-second day, all was healed except for a small gray spot that Pierre blamed on a deeper injury. Although the Curies and Becquerel had problems with their hands and fingers, they dismissed the burns as minor problems that had to be dealt with. In fact, it seemed that radium would be actually useful. They postulated that by destroying diseased cells, radium could cure certain forms of cancer. It seemed that radium might be a medical miracle.

Although both Curies worked very hard, they took time off for a vacation each summer from 1900 through 1903. Marie was especially concerned about Irène's health and felt that these summers away from Paris were essential to her well-being. When they returned to Paris each year, both parents suffered from extreme exhaustion but did not blame radium for their symptoms. In May 1902, Marie's beloved father, Wladyslaw, died. Before he became ill, Marie and Pierre had visited him in Warsaw several times. In 1899 the entire family (Wladyslaw, Bronia, Kazimierz, Helena, and Józef) were reunited in Zakopane in the

Carpathian Mountains where the Dluskis were building a new modern tuberculosis sanitorium. Shortly after that visit, Marie's father was struck by a truck and suffered a debilitating fracture. Although he recovered somewhat, shortly thereafter he had a gallbladder attack and had surgery to remove large gallstones. As soon as she heard of his illness, Marie took a train to Warsaw, but en route she learned that he had died. His death affected her greatly, particularly because she was not able to be at his side when he died. She was somewhat comforted in the knowledge that he who had wanted to do scientific work as a young man was proud of the scientific success of his daughter.

NOTES

1. Marie Curie, "Autobiographical Notes," in *Pierre Curie,* trans. Charlotte and Vernon Kellogg (New York: Macmillan, 1923), p. 180.
2. Ibid., p. 181.
3. Ibid., p. 182.
4. Robert Reid, *Marie Curie* (London: Collins, 1974), p. 86.
5. Marie Curie, "Autobiographical Notes," p. 184.
6. Eve Curie, *Madame Curie: A Biography* (Garden City, NY: Doubleday, Doran & Co., 1938), p. 169.
7. Reid, *Marie Curie*, p. 96.
8. Marie Curie, "Autobiographical Notes," p. 186.
9. M. P. Curie, Mme. P. Curie, and M. G. Bémont, presented by M. Becquerel, "Sur une nouvelle substance fortement radio-active, contenue dans la pechblende," *Comptes Rendus* 127 (1898): 1215–17. The English translation of the title is "On a New, Strongly Radioactive Substance, Contained in Pitchblende." The translation is by Henry A. Boorse and Lloyd Motz, eds., *The World of the Atom,* vol. 1 (New York: Basic Books, 1966).
10. M. P. Curie, Mme. P. Curie, and M. G. Bémont, presented by M. Becquerel, "Sur une nouvelle substance fortement radio-active," pp. 1215–17.
11. Marie and Pierre Curie, "Les nouvelles substances radioactives et les rayons qu'elles emettent," rapport presente au Congres international de physique (Paris: Gauthier-Villars, 1900), in Susan Quinn, *Marie Curie: A Life* (New York: Simon and Schuster, 1995), p. 450.
12. Ernest Rutherford, "Mme. Curie," *Nature,* July 21, 1934, pp. 90–91.
13. Helena M. Pycior, "Pierre Curie and 'His Eminent Collaborator Mme.

Curie,'" in *Creative Couples in the Sciences,* ed. Helena M. Pycior, Nancy G. Slack, and Pnina G. Abir-Am (New Brunswick, NJ: Rutgers University Press, 1996), p. 39.

14. Eve Curie, *Madame Curie*, p. 162.
15. Ibid., p. 163.
16. Ibid.
17. Ibid.
18. Marie Curie, "Autobiographical Notes," p. 179.
19. Marie Curie, *Pierre Curie*, trans. Charlotte and Vernon Kellogg (New York: Macmillan, 1923), pp. 108–109.
20. Eve Curie, *Madame Curie*, p. 184.
21. Ibid., p. 186.
22. Quinn, *Marie Curie: A Life*, p. 179.

Chapter 6

A YEAR OF CONTRASTS: GOOD NEWS, BAD NEWS

The year 1903 was one of contrast for the Curies. As discussed in the last chapter, it was the year in which Marie's beloved father died. Yet it was also the year in which she brilliantly defended her doctoral thesis and became the first French woman to earn a doctorate. An examining committee comprised of two physicists and one chemist (including two future Nobel Prize winners) declared that she had defended her thesis with "distinction." But this year, more than most, seemed to have more than its share of ups and downs for the Curies. Pierre, accompanied by Marie, made a trip to London to present an invited lecture at the Royal Institution. Pierre's health had become increasingly fragile during the previous year. Immediately before the lecture Pierre became so ill that he even had difficulty dressing himself. However, once he started to speak he seemed to revive. His talk was well received and his party tricks with the radium that had probably caused his sickness were especially appreciated. The lecture that could have been a disaster because of his health turned out to be a great success. During one demonstration, he spilled a minuscule quantity of radium; fifty years later the level of radioactivity in the building was sufficient to require decontamination. While Pierre was lecturing, Marie, who had done much of the work that he was describing, sat in the audience, giving the audience the impression that Pierre was the more important scientist. However, Pierre himself was careful to acknowledge his wife's essential role in their collaborative work.

While they were in England, Marie and Pierre met the elite of British science: Sir William Crookes (1832–1919), Lord Rayleigh (1842–1919), Sir Oliver Lodge (1851–1940), H. E. Lankester (1814–

1874), Ray Lankester (1847–1929), and, of course, Lord Kelvin. Although Kelvin never believed that radium was a new element, he was very kind to the young couple. He showed them through his laboratory and seemed greatly interested in their research. During this trip they were hosted by the astronomers Sir William (1824–1910) and Lady Margaret (1848–1915) Huggins. This experience was especially meaningful to Marie, because the Hugginses were also a collaborative scientific team. Their collaboration was different from the Curies' because Margaret had no formal training, although she had as a child exhibited a passionate interest in astronomy. It still was gratifying to find another couple working so successfully together. Margaret developed many skills and the couple's common interest in spectroscopy that brought them together persisted throughout their careers. Although Margaret was usually characterized as William's assistant, he later recognized her very real contributions. In a paper on the Orion Nebula he noted, "I have added the name of Mrs. Huggins to the title of the papers, because she has not only assisted generally in the work, but has repeated independently the delicate observations made by the eye."[1] Marie was very impressed by the way Margaret and William worked together.

In this same year Marie lost a child, born prematurely after one of their bicycle rides. During her pregnancy, she had been exposed to extremely high doses of radiation, although she did not relate it to the miscarriage. She wrote to Bronia on August 25, 1903, about her disappointment at the baby's death. She explained that the little girl was still living when born and that she "had wanted it badly."[2] Friends had previously urged Marie to take better care of herself. A colleague, Georges Sagnac, had written to Pierre during the previous spring, berating the couple for not eating properly. Sagnac's recipe for a cure involved regular mealtimes without reading or talking physics. Looking back on this and other warnings as well as what she knew herself, Marie decided that her lifestyle was to blame for the miscarriage. This self-blame caused an almost debilitating guilt.

Another calamity affected Maria greatly. Bronia lost her five-year-old son later that same year to meningitis. She wrote to her brother, Józef, about the tragedy. The little boy who died had been the picture of health, and "if, in spite of every care, one can lose a child like that, how can one hope to keep the others and bring them up?" Fearing a similar fate for Irène, she continued, "I can no longer look at my little girl without trembling with terror."[3]

After her miscarriage and the death of her nephew, Maria was ill during the remainder of the summer and well into the fall. While she was convalescing, they took a long vacation on the Ile d'Oléron. By the end of September Marie pronounced herself cured, although the doctors found her anemic. Anemia can be brought on by exposure to radium, and it was several months more before she had the strength to go back to work.

On the positive side, several honors and prizes came to Marie and Pierre this year. Some of their French colleagues began to recognize the importance of their work. Previously (1902) the Académie des Sciences of the Institute had awarded the Curies twenty thousand francs for isolating radium. Then in 1903 they presented the coveted Gegner Prize for the third time to Marie for scientific promise. During the same year, Pierre won the ten-thousand-franc biannual Prix La Caze. In November 1903, the Curies received the Davy Medal presented by the Royal Society of London while Marie was still recuperating.

In the June 25, 1903, issue of an American popular journal, the *Independent,* Marie Curie described her investigations on radioactive substances. The editor established her credibility as a scientist by explaining that "she has published two or three works on physical subjects" and would soon be defending her thesis before the Sorbonne (University of Paris). After the defense, he asserted, she would have her doctor's degree, "the highest degree given in France." The editor claimed that "this is the first magazine article, we believe, that has appeared on the radio-active elements from either Madame Curie or her husband."[4] Marie's article was among others on topics popular at the time, including "Servia: Its Present and Its Future," "The Hotel Martha Washington," "Women in Church Work," and "Latin-American Revolutions." Obviously, the editor recognized the importance of the Curies' work and saw fit to publish it in a form that the general public could understand. What was so amazing, he reported, was that "when the new wonders of radium were announced to the world a few months ago . . . there was a rumor that a woman was associated with the remarkable discovery." Some people, he noted, were reluctant to believe that Marie had an important part in the discovery. But "a perusal of the following pages will show that Professor Curie is rather the helpmate of his wife in this magnificent piece of scientific work."[5] Marie Curie demonstrated to the American public in this short article that she could succinctly present her ideas and discoveries in such a way that the general educated public could understand them.

THE NOBEL PRIZE, 1903

The *New York Times*'s headline for December 11, 1903, listed the Nobel recipients for the year, noting that the prize for physics was divided between Henri Becquerel of Norway and the Curies of Paris. The article noted that the Curies were the best known of the prize recipients. Because they were not profiting financially from the work, "their admirers throughout the world will be delighted to hear of this windfall for them."[6]

Even though Marie's accomplishments were recognized from as far away as the United States, she almost missed out on the most prestigious award of them all, the Nobel Prize. In December 1903, the Curies and Henri Becquerel were jointly awarded the Nobel Prize for physics. Pierre Curie was nominated by four members of the Académie to share the Nobel Prize for physics with Henri Becquerel, completely leaving Marie out. One of the prize committee members, Gösta Mittag-Leffler, a great supporter of women scientists, wrote to Pierre and explained that only he and Becquerel were to be nominated for the prize. Pierre answered Mittag-Leffler's letter stating, "if it is true that one is seriously thinking about me [for the prize], I very much wish to be considered together with Madame Curie with respect to our research on radioactive bodies."[7] However, since Marie Curie had not been nominated for the 1903 prize it seemed as if she would be ineligible. The situation was saved because Marie had received two votes for the previous year's prize. By allowing one of these nominations to be valid for 1903 she was permitted to share the prize with her husband and Becquerel. As the Swedish Academy of Sciences discussed the Curies' nomination, they changed their original intent, which was to award the Curies the prize in physics for their discovery of spontaneously radioactive elements. The chemists, however, objected because they wanted to leave the door open for the Curies to receive a second prize, in chemistry, for the discovery of radium. Thus, they decided to give the Curies the prize in physics in 1903 "for their joint researches on the radiation phenomenon discovered by Professor Henri Becquerel."[8] There was a tacit understanding that a prize in chemistry might eventually be forthcoming.

As a woman Nobel laureate, Marie Curie was an oddity in 1903 and would still be so today. In three scientific disciplines (physics, chemistry, and physiology and medicine) women Nobel Prize winners are still very scarce. From Marie Curie's 1903 prize in physics through Christiane-Nusslein-Volhard's prize in physiology and medicine in 1995, there have been only eleven women Nobel Prize winners in the sciences, compared to over four hundred men from 1901 (when the first prize was given) through 2000. Marie Curie was more acceptable because she worked in collaboration with her husband.

In the prize presentation speech given by H. R. Törnebladh, the president of the Royal Swedish Academy of Sciences, he noted that the success of the Curies illustrated an old proverb, "union is strength." He also quoted a biblical passage stating, "it is not good that the man should be alone; I will make him a help meet for him." He continued by noting "this learned couple represent a team of differing nationalities, a happy omen for mankind joining forces in the development of science."[9]

The prize forever destroyed the Curies' voluntary isolation. The prize catapulted them into fame and forced this shy retiring couple into the unasked for and unwanted limelight. In a letter to Georges Gouy (1854–1926) on January 22, 1904, Pierre excused himself for not writing sooner "because of the stupid life I am living now." He continued:

> You have seen this sudden fad for radium. This has brought us all the advantages of a moment of popularity; we have been pursued by the journalists and photographers of every country on earth; they have even gone so far as to reproduce my daughter's conversation with her nurse and to describe the black-and-white cat we have at home. Then we have received letters and visits from all the eccentrics. . . . We have had a large number of requests for money. . . . With all this, there is

> not a moment of tranquility in the laboratory, and a voluminous correspondence to be sent off every night. On this regime I can feel myself being overwhelmed by brute stupidity.[10]

Marie wrote to her brother that "our life has been altogether spoiled by honors and fame" and to her cousin Henrietta that "our peaceful and laborious existence is completely disorganized: I do not know if it will ever regain its equilibrium."[11]

Becquerel went to Stockholm to receive his award, but the Curies, who were both unwell, blamed their teaching schedules as the reason for their absence. It was not until June 1905 that the Curies were able to travel to Sweden, where Pierre, dressed in formal clothes and trembling with shyness, gave the lecture required of Nobel recipients. His voice initially quavered, but once he began to explain their discoveries he captivated the audience.

The fact that Marie shared the prize with two men is taken for granted today. But when the Curies were doing their research it was widely assumed that Marie was merely Pierre's assistant. Both Pierre and Marie wanted to be certain that the male-dominated scientific community gave Marie the recognition that she deserved. Pierre was modest and unassuming about his own accomplishments. Collaboration was second nature to him, because his earliest work in science had been done with his brother Jacques. He enjoyed sharing ideas and, as Marie wrote in her biography of Pierre, his years of collaborating with Jacques were both "happy and fruitful." Thus Pierre was prepared to see Marie as an equal partner. Marie explained that "their devotion and their common interest in science were to them both a stimulant and a support."[12] Even more important than Pierre's attitude was Marie's own self-confidence. Because of her belief in her own abilities she did not hesitate to publish under her own name the works for which she alone deserved the credit independently. She also recorded the experimental results in her notebooks that related to her own work. In 1898 she published a note under her own name announcing the discovery of thorium's radioactivity as well as the hypothesis that pitchblende contained a new element. Concerning the latter conclusion, she wrote that since two minerals found in uranium ore, pitchblende and chalcolite were more active than uranium itself this "leads one to believe that these minerals may contain an element much more active than uranium."[13]

Her later accounts of radioactivity always mentioned her independent contributions. She was also careful to credit Pierre with his independent work as well as his publications that resulted from collaborating with others. In 1898 they submitted one of many joint papers to the French academy. This one was on polonium. In this, as in all of their joint papers, they were careful to credit their individual contributions.[14] This publication policy effectively enhanced the Curies' reputation as a collaborative couple and helped to solidify Marie Curie's reputation as an independent scientist. Still, when Pierre and Marie collaborated, Pierre took over the role as the lead scientist. His name appeared first on their joint papers. Although the couple was concerned about assuring that Marie was recognized for her work, they did not openly challenge the early twentieth-century ideas that supported the male scientist over the female.

CUTTING THE UNCUTTABLE

Even though Marie Curie had postulated that radioactivity was atomic in nature, she still did not know what caused a substance to be radioactive. Others were equally perplexed. Once the explanation was finally understood, scientists used her original idea of radioactivity as an atomic quality to build a new physics. However, the idea of the atom was not new at all. Ever since the time of the ancient Greeks (fifth century BCE), people proposed two ways of looking at matter. They either assumed that the universe was completely full—made up of a continuum with no empty spaces—or that it was composed of tiny indivisible particles moving in space (a void). The latter idea was known as atomism. The very word atom comes from the Greek *atomos*, meaning uncut. Some of the evidence from the work on radiation implied that this ultimate particle could be transformed into even smaller units. Although the grand old man of nineteenth-century physics, William Thomson, Lord Kelvin (1824–1907), insisted until the end of his life that the atom was indestructible, others were evolving an entirely new theory of matter. They reluctantly concluded that the atom was not the indivisible particle hitherto assumed. The first "cut" in the uncuttable atom was made by Joseph John Thomson (1856–1940). From an experiment deflecting cathode rays, he postulated that these rays were streams of particles much smaller than atoms. He concluded that these very light particles were universal constituents of matter. Although Thomson called these particles "corpuscles," the term "electron" that previously had been invented by G. J. Stoney (1826–1911) ultimately became accepted for this negatively charged particle. At this point an atom seemed to consist of negative electrons plus the rest of the atom.

Another scientist, the New Zealand–born Canadian scientist Ernest Rutherford (1871–1937), contrived an experiment that resulted in an additional cut in the uncuttable atom. Rutherford had worked under J. J. Thomson at Cambridge. Rutherford went to McGill University in Canada where he devised an experiment that led to the notion that the atom could be cut into yet another part. At his suggestion, his colleagues Hans Geiger (1882–1945) and Ernest Marsden (1889–1970) shot alpha (positively charged) particles at a thin sheet of gold, assuming that the particles would go straight through the foil with little deflection. According to the accepted theory (known as the plum pudding model of the atom), the negative electrons (the plums or raisins) would be spread evenly throughout the positive matrix (the pudding). They were surprised to find that although most of the positive alpha particles went straight through the foil (98 percent), a small percentage were deflected at large angles (about 2 percent), and .01 percent bounced off the gold foil. Since alpha particles have about eight thousand times the mass of an electron, it became clear that very strong forces were necessary to deflect the particles. Rutherford interpreted these results to mean that most of the mass of an atom was concentrated into a compact positive nucleus with electrons occupying most of the atom's space. Accepting this model meant that most of the atom was space—very different from the plum pudding model.

Later, the French scientist Paul Villard (1860–1934) showed that radioactive substances produced a third set of penetrating radiations, which were later called gamma rays and had a neutral charge. These discoveries made it apparent that the indestructibility of the atom was a myth. The uncut atom was now hypothesized to be composed of three different particles, a positive proton, a neutral neutron, and a negative electron. By 1913, a picture of the atom was conceived that resembled the one that is accepted today with a nucleus composed of protons and neutrons surrounded by electrons. Niels Bohr combined the quantum theory with Rutherford's model of the atom to provide a model of the atom that is familiar to us. In spite of the fact that today the atom is still being cut into additional subatomic particles, scientists still use the ancient term atom (uncut). We know that neither the ancient Greeks nor modern scientists have ever directly seen an atom (by the human eye). Nevertheless, scientists are confident of its existence. This brings up the question of the nature of science. If science is an attempt to explain natural phenomena by creating theories that agree with observations, then what we know about the existence of the atom makes sense. We create a theory to explain the observational or experimental results. If the theory fails this test, then scientists may attempt to resurrect it by adding additional postulates or, if all else fails, replace it by a new theory. In the case of the atom, in order to explain the increasingly complex observations, it has been necessary to cut the uncuttable atom into increasing numbers of subatomic particles.

One of the Curies' early observations needing an explanation was the observation that radium gave out heat in large enough amounts to be measured by simple laboratory techniques. They asked where the heat energy came from. This observation seemed to break a basic law of physics, the law of conservation of energy. This law states that energy can neither be created nor destroyed. If the heat energy was not being created, where did it come from? Marie Curie proposed two possible explanations—either the radioactive substances were borrowing energy from an external source and then releasing it, or that the radium was the source of the energy itself. Rutherford was familiar with the Curies' results and their theoretical speculations. He blew air across samples of thorium and found that he could collect a radioactive gas, which he called thorium emanation. He found that it diminished over time. Others found different radioactive substances also produced emanations. If an emanation came into contact with a substance that was not radioactive, the radioactivity of the emanation would be transferred to the new substance. We have already seen that collaboration with a colleague can be a potent way to solve a scientific problem. Teamwork in the case of the Curies' research was essential.

Another case in point is the collaboration between Rutherford, the physicist, and Frederick Soddy (1877–1956), the chemist. Working together they showed that radioactive elements by giving out either alpha or beta rays would break down into intermediate elements. Each of the intermediate elements broke down at a specific rate so that half of any quantity disappeared in a fixed amount of time. Rutherford called this time the half-life of the substance. By building on the work of the Curies and others, they had made a discovery that explained the nature of radioactivity. They had accomplished something that alchemists had attempted for a thousand years—to transmute (change) one element into another. The early alchemists were convinced that with the proper techniques, base metals such as lead could be transmuted into beautiful and valuable gold. In Rutherford and Soddy's work, putting the substances through elaborate distillations, water baths, and dung baths, as the early alchemists did, proved unnecessary. The radioactive materials could transmute all by themselves. Since alchemy was in disrepute by this time, Rutherford and Soddy hesitated to use the term transmutation.

AFTER THE PRIZE (1904–1905)

Although the Nobel Prize money relieved the Curies of their most burdensome financial problems, the next two years had both positive and negative aspects. Both Pierre and Marie suffered physically from the effects of radiation, although they still did not recognize why they felt so ill. Pierre felt underappreciated for in spite of the Nobel Prize he still did not have a coveted professorship at the Sorbonne. When he was told that there was a possibility that he would finally get a chair, he was concerned that he would again be disappointed. This time, however, he was successful. The French Parliament created a new professorship especially for him. It included a laboratory and a small support staff with Marie as laboratory chief. For the first time in her career, she had official rights in Pierre's laboratory. She was appointed director of

Pierre's research laboratory beginning on November 1, 1904. She received an annual salary of 2,400 francs.

Although Marie pined for the child she had miscarried in the fifth month, she soon became pregnant again. While pregnant, she temporarily gave up her teaching post at Sèvres but continued her research. Pregnancy was always difficult for Marie, and this time was no exception. Although she was exhausted after the birth, Eve Denise, a perfect baby girl, was born on December 6, 1905. The "perfect baby girl" was not an easy baby to care for. In a letter to Józef, Marie reported that Eve was seldom interested in sleeping and protested vehemently if left in her cradle by herself. She noted that Eve and Irène were very different physically. Whereas Eve had dark hair and blue eyes, Irène had light hair and green-brown eyes.

After Eve's birth, Marie returned to her teaching post at Sèvres, although the prize money made it unnecessary. She enjoyed teaching young women and associating with her colleagues there, especially with Paul Langevin, a physicist whom she and Pierre knew under other circumstances and who taught at Sèvres. Inexperienced when she first began to teach, Marie was ridiculed by her students because of both her accent and her style of teaching. After examining her own teaching methods, she came to the conclusion that the students needed hands-on experience rather than abstract lectures. The happy result of her laboratory-based classes resulted in her becoming one of the most popular teachers at the school.

Although her life was very busy with home, children, husband, teaching, and research (not to mention dealing with the press and those who were fascinated by her prize), Marie still found more time than previously to socialize with friends. Both Curies still attended the salon evenings of Marguerite Borel, the wife of the mathematician Émile Borel, who sometimes found the couple intimidating. Although Marie remained careful with money, after the prize she bought new clothes for herself, Pierre, and the children. The American dancer Loie Fuller, who made a huge splash in Paris by dancing with veils illuminated by colored lights, presented a show especially for the Curies at their house. There were many less-formal occasions at the Curie house where scientists, artists, and authors discussed a variety of ideas. Their close neighbors were Jean and Henriette Perrin. Jean was a scientist and political activist. Henriette Perrin was Marie's closest woman friend at this time.

Once again, Pierre was bullied into becoming a candidate for the French Academy of Sciences. Although the vote was very close, this time he was successful. The narrow win did not help his ego. In a 1905 letter to Georges Gouy he wrote:

> I find myself in the Academy without having desired to be there and without the Academy's desire to have me. I only made one round of visits, leaving cards on the absent ones, and everybody told me it was agreed that I would have fifty votes. That's probably why I nearly didn't get in.[15]

Later he again wrote to Gouy complaining, "I have not yet discovered what is the use of the Academy."[16]

Neither Marie nor Pierre agreed to patent commercial radium production. They thought it crass to use science for commercial purposes. Whenever they were asked for information on the radium separation process, they gave it freely. Their view was that scientists were supposed to share ideas and techniques with others. Others managed to benefit financially from their generosity. By commercializing the process they invented, less idealistic entrepreneurs were getting rich, while the Curies were forced to ask a rich benefactor for money to continue their research. By 1906, Pierre's sickness had gotten decidedly worse. It is ironic that the last published paper in his lifetime (1904), written with two medical colleagues, was concerned with the experimental effects of radioactive emanations on mice and guinea pigs. When they performed postmortems on the animals, they found profound pulmonary congestion and distortions in the white blood cells (leucocytes) that protect the body from infectious diseases. It was obvious that radium gas had devastating effects on the animals. But they did not seem to apply this information to illness in laboratory workers or to themselves.

NOTES

1. William and Margaret Huggins, "On the Spectrum, Visible and Photographic of the Great Nebula in Orion," *Proceedings of the Royal Society* 46 (1889): 40–61.

2. Susan Quinn, *Marie Curie: A Life* (New York: Simon and Schuster, 1995), p. 184.

3. Eve Curie, *Madame Curie: A Biography* (Garden City, NY: Doubleday, Doran & Co., 1938), p. 191.

4. Madame Sklodowska-Curie, "The Radio-Active Elements," *Independent: A Weekly Magazine* 15 (June 25, 1903): 1498–1501.

5. Ibid.

6. "Nobel Prizes Awarded," *New York Times*, December 11, 1903, p. 8, col. 6.

7. Sharon Bertsch McGrayne, *Nobel Prize Women in Science: Their Lives, Struggles, and Momentous Discoveries*, 2nd ed. (Washington, DC: Joseph Henry Press, 1998), p. 25.

8. Ibid., p. 26.

9. H. R. Törnebladh, president of the Royal Swedish Academy of Sciences, "Presentation Speech. Nobel Prize in Physics, 1903," in *Nobel Lectures in Physics: 1901–1921* (Amsterdam: Elsevier, for the Nobel Foundation, 1967), pp. 50–51.

10. Eve Curie, *Madame Curie*, p. 216.

11. Ibid., p. 217.

12. Marie Curie, *Pierre Curie*, trans. Charlotte and Vernon Kellogg (New York: Macmillan, 1923), p. 48.

13. Marie Curie, "Rayons émis par les composés de l'uranium et du thorium," *Comptes rendus de l'Académie des Sciences* 126 (1898): 1101–102.

14. Pierre and Marie Curie, "Sur une substance nouvelle radio-active, contenue dans la pechblende," *Comptes Rendus* 127 (1898): 178–80.

15. Eve Curie, *Madame Curie*, p. 235.

16. Ibid.

Chapter 7

"PIERRE IS DEAD? DEAD? ABSOLUTELY DEAD?"

In April 1906 Pierre joined Marie and the children for a brief holiday in the country. Even though spring in France can be cold and raw, the weather was fine during these days. Although Pierre's fatigue concerned Marie, the family basically had a pleasant relaxing time away from the pressures of Paris. Pierre returned to the city on April 14, but Marie and the girls stayed on. After the weather turned cold and rainy they returned on April 16. On that same night the Curies attended a dinner meeting of the physics society. One of the subjects of conversation at the dinner had to do with spiritualism. During the late nineteenth and early twentieth centuries, many scientists dabbled in spiritualism. Both Pierre and Marie had attended séances given by the medium Eusapia Palladino. Pierre was particularly interested, and was disturbed that he could find no obvious way to discredit the ghosts that she produced. He was not the only scientist who was fascinated with the spirit world. Other nineteenth-century scientists such as Sir William Crooks, Alfred Russel Wallace, and Charles Robert Richet (1850–1935) also were believers. On April 17, Pierre attended another meeting of the physics society and put forth his thoughts on another new passion, the teaching of science. He was excited when his ideas were accepted and he was elected vice president of the new organization.

The following day, Thursday, April 19, Pierre left his laboratory and walked in the chilling rain to a meeting of the new group. After the meeting was over he felt exhilarated. His thoughts were on the people who he had met and the way in which they had appreciated what he had to offer. The meeting broke up shortly after 2:00 in the afternoon. Because of the cold drizzle, Pierre put up his umbrella and headed toward the Institut de France. As he neared his destination, both the rain and the traffic became heavy. He had been walking behind a cab, which provided him with some shelter. As he came to a very busy intersection he abandoned the cab and absentmindedly stepped into the path of a horse that was pulling a wagon. At the same moment the wagon was passing the cab. The space between the cab and the wagon shrunk dangerously. Pierre, whose mind was on anything but walking in traffic, was surprised at the narrowing space. When it seemed that he would be crushed, he caught the horse by the chest in an attempt to hang on. The horse suddenly reared, and Pierre slipped on the wet pavement as the crowd shouted to the driver to stop. Although the driver pulled on the reins, the team of horses continued on. Pierre was on the ground but unhurt and lay very still at the feet of the horses as the two front wheels of the wagon moved by without touching him. But salvation was brief, for the wagon's left rear wheel hit and smashed Pierre's head. A crowd quickly gathered. Some of the gawkers tried to entice passing cab drivers to take the body to the police prefecture, but the drivers refused because they feared that the bloody body would stain their upholstery. Eventually two men brought a stretcher and Pierre was carried to a nearby police station. At the station they examined his papers and realized who he was. Although his head was crushed, Pierre's face was recognizable, and one of his laboratory assistants identified the corpse. When the crowd realized that the victim was their Nobel Prize–winning scientist, it turned on the driver of the wagon and the police were forced to intervene in order to protect him.

It was decided that Paul Appell, Pierre's senior colleague and dean of the faculty of science, and Jean Perrin, the Curies' next-door neighbor and dear friend, would tell Marie about the accident. However, when they arrived at the house Marie was not yet home. Dr. Eugène Curie, Pierre's father, was alone in the house. When he saw the looks on the faces of Appell and Perrin, his first comment was, "My son is dead." He was heartbroken and through his tears accused Pierre of absentmindedness and uttered reproachfully, "What was he dreaming of this time?"[1] Was the accident actually caused by daydreaming? Did his umbrella keep him from seeing the vehicles? Or did the slick street, combined with his illness, make him less sure-footed than usual? Although nobody knows exactly what happened, it seems most likely that it was a combination of all of the above.

Above: Wladyslaw Sklodowski and daughters Maria, Bronia, and Helena (1890). Image courtesy of ACJC-Curie and Joliot-Curie fund.

Left: Marie Curie as a young woman. Image courtesy of the History of Science Collections, University of Oklahoma Libraries.

Above: Paris in 1902, looking toward Notre Dame from the Eiffel Tower. Image courtesy of the History of Science Collections, University of Oklahoma Libraries.

Right: Marie and Pierre Curie with their new safety bicycles, 1895. Image courtesy of ACJC-Curie and Joliot-Curie fund.

Left: The Eiffel Tower in 1902. Image courtesy of the History of Science Collections, University of Oklahoma Libraries.

Below: Pierre Curie and his electroscope (1923). Image courtesy of ACJC-Curie and Joliot-Curie fund.

Pierre and Marie Curie in their laboratory. Image courtesy of the History of Science Collections, University of Oklahoma Libraries.

The radium shed. Image courtesy of the History of Science Collections, University of Oklahoma Libraries.

A Vanity Fair *caricature of Marie and Pierre Curie with radium. Image courtesy of the History of Science Collections, University of Oklahoma Libraries.*

Marie with her children in 1908. Image courtesy of ACJC-Curie and Joliot-Curie fund.

Paul Langevin. Image courtesy of ACJC-Curie and Joliot-Curie fund.

Marie Curie, Albert Einstein, and others at the 1927 Solvay Congress in Brussels, Belgium. Image courtesy of the History of Science Collections, University of Oklahoma Libraries.

Above: Marie Curie and President Harding as she was presented with the gram of radium in 1921. Image courtesy of the History of Science Collections, University of Oklahoma Libraries.

Left: Marie Curie in 1921, photograph given to Vassar College. Image courtesy of the History of Science Collections, University of Oklahoma Libraries.

Right: Marie and Irène Curie, working together. Image courtesy of the History of Science Collections, University of Oklahoma Libraries.

Below: Paris in 1902, looking toward the Pantheon, where Marie is now buried, from the Eiffel Tower. Image courtesy of the History of Science Collections, University of Oklahoma Libraries.

When Marie arrived home at six o'clock, she knew from the demeanor of her friends that something was very wrong. Paul Appell reiterated the facts. Marie remained motionless and listened numbly to what he had to say. She did not cry; when she finally spoke it was to say, "Pierre is dead? Dead? Absolutely dead?"[2] As the drama unfolded, she still could not really grasp that Pierre, her lover, companion, the father of her children, and her scientific collaborator, was gone. The stress of dealing with Irène was too much, so she asked Mme. Perrin to care for her for several days, and gave herself up to mourning. Eve remained at home cared for by others. She could not bring herself to explain to Irène that Pierre was dead, and when she and Irène talked across the fence Marie simply told her that her father had hurt himself badly in the head and needed rest.

On the evening of Pierre's death, friends brought Marie the few articles found in his pockets: fountain pen, keys, wallet, and a watch. She sent a terse telegram to her family in Poland stating simply that Pierre was dead from an accident. André Debierne went to the police station to retrieve the body of his friend and brought it to Marie. Left alone with her husband, she kissed his face and refused to stay in a room away from Pierre while they washed and dressed the body. When Jacques Curie arrived, she was finally able to express her grief and broke down in sobs. Jacques' presence was a great comfort. She later noted that they reread the old letters and what remained of his journal.

Pierre Curie's death was noted immediately in the United States. In a headline the *New York Times* reported, "Prof. Curie Killed in a Paris Street." The subheading read "The Discoverer of Radium Run Over by a Wagon." An additional subhead read "Success Followed Early Hardship—Curie Was Greatly Aided by Mme. Curie." Further down in the obituary notice, the article again mentioned Marie not as Pierre's collaborator but only as his assistant. "In his researches he was aided by Marie Sklodowska, a Pole, who was born at Warsaw, in 1868." The author of the article seemed convinced that a woman could only serve as an assistant to her husband, not as a full partner. The article contained another mistake, mentioning that Professor Curie left only one child, a nine-year-old daughter.[3]

Keeping a journal provided Marie with the therapy that helped her deal with the tragedy. She addressed the journal entries to Pierre. Begun on April 30, 1906, she wrote, "Dear Pierre who I will never more see here, I want to speak to you in the silence of this laboratory, where I never thought I would have to live without you."[4] The section describing Pierre's interment was especially poignant. She wrote that "your coffin was closed and I could see you no more. I didn't allow them to cover it with the horrible black cloth. I covered it with flowers and I sat beside it."

After they took the body to Sceaux for burial, she expressed her horror at Pierre being placed in a deep hole. "They filled the grave and put sheaves of flowers on it. Everything is over, Pierre is sleeping his last sleep beneath the earth; it is the end of everything, everything, everything."[5]

The journal also expressed minor problems in their marriage. Small tensions between the two emerged as Marie chided Pierre for working too hard and not spending enough time with the family. But she also recalled the wonderful vacations that they had taken together. As she reminisced, she would be jerked back to reality. These times were at an end. Pierre was dead. Biographer Susan Quinn reports that an entire page was torn out of the journal, but speculates that it was either Marie or someone in Marie's family who censored it.

The day after the burial, Marie finally explained to Irène what had happened. Eight-year-old Irène was playing with her friend Aline Perrin when her mother decided that the time had come to tell Irène. At first the news seemed to wash over Irène's head. She did not appear to understand and went back to playing with Aline. However, after Marie left, she burst into tears and Henriette Perrin, Jean's wife, took her back to her mother. Marie wrote, "She cried a great deal at home, and then she went off to her little friends to forget. She did not ask for any detail and at first was afraid to speak of her father." Józef and Bronia came to lend their support, but the depth of Marie's despair frightened them as well as Dr. Curie and Jacques. In her journal she wrote, "In the street I walk as if hypnotized, without attending to anything. I shall not kill myself. I have not even the desire for suicide. But among all these vehicles is there not one to make me share the fate of my beloved?"[6]

Of the tributes to Pierre, none was more poignant than that of his close friend and colleague Paul Langevin, published in the *Revue du mois*. He observed that just as Pierre's life was improving, and he could spend all of his time in his precious laboratory without teaching responsibilities, he was cruelly killed.

Marie got financial advice from Georges Gouy, who advised her not to mention her own radium when she made an inventory of Pierre's laboratory for the Faculty. He explained that radium had become so valuable that she might have to pay death duties if she reported it. After suggesting that she solicit the advice of a competent businessman to help with any issues the radium might bring up, he admonished her to think of the futures of Irène and Eve, even if she herself was uninterested in personal gain. Against the advice of those who thought she should keep the radium (worth more than a million gold francs), she gave it to the laboratory instead.

After several weeks passed, new questions arose. What would happen to Pierre's research? Who would take over his teaching at the Sorbonne? The university agreed to give Marie, as Pierre's widow, a pension, but she refused to accept it. She flatly stated that she was too young to accept a pension; she could support herself and her children. The *New York Times* reported on the pension, apparently unaware of Marie's decision to decline it. The newspaper reported that the "Council of Ministers has decided to have the Minister of Education introduce a bill in the Chamber of Deputies for a pension for the widow and children of Prof. Curie, the discoverer of radium, who was killed in Paris last Thursday by being run over by a wagon in the Place Dauphine."[7] This US newspaper was still convinced that Pierre had discovered radium.

Since Marie was still in a dazed state and was not prepared to make decisions about the future, her family and friends took the initiative. They informed the dean that Marie was the only French physicist competent to succeed Pierre. Although no woman had ever held such a position, the council of the Faculty of Science unanimously decided to offer her an assistant professorship. She also was given the chair especially created for Pierre, which he had only occupied for eighteen months. Almost a month after Pierre's death Marie reported in her journal that she had been officially named Pierre's successor. The entries in her journal stopped between June and November, although she continued to write progress reports on the children in her notebook.

The time when Marie decided to recognize that Pierre was actually gone and that she would return to living occurred in the middle of June. Eve recounts the evening when the decision was made. Marie motioned to Bronia, who had remained in Paris after Pierre's death, to follow her to her bedroom. Removing a package wrapped in waterproof paper from the cupboard she asked Bronia's help. She untied the string and opened the parcel and released a white cloth that wrapped a grotesque collection of bloody clothes and dried mud. These were the clothes that Pierre had worn when the wagon struck him. Marie took a pair of scissors and began to cut up the coat, throwing the pieces one by one into the blazing fire. She stopped when some fragments of brain tissue appeared, dissolved into tears, and kissed them passionately until Bronia grabbed the scissors and continued cutting and burning the remains of the clothing. After this Marie made the decision to concentrate on her remaining family and her science, although joylessly.

During much of the summer and early fall, Marie prepared to teach Pierre's course at the Sorbonne. Since she was the first woman to teach there, she was well aware that many people would be watching her lectures with interest. In many ways, her success would reassure the world that a woman could be a successful professor at one of the most famous universities in the world, a reassurance some did not want. Several hundred people gathered for her first lecture, many hoping for some drama. Perhaps she would break down in tears while giving a tribute to her late husband. A few others secretly hoped that she would fail, confirming their prejudices that a woman should not hold such a prestigious teaching and research chair. Neither group got what it wanted. To a thundering ovation, Marie stared straight ahead and calmly began her lecture: "When one considers the progress that has been made in physics in the past ten years, one is surprised at the advance that has taken place in our ideas concerning electricity and matter."[8] These clear unemotional words affected the audience more than hysterical weeping would have done. Many of the listeners felt tears slipping down their cheeks, for Marie had resumed the course at the exact sentence where Pierre had left it. After completing the lecture, she left the hall quickly.

In her biography of Pierre, Marie quotes some of the eulogies of Pierre written by his friends and colleagues. The following two examples indicate the love and respect of two of his fellow scientists. His student and close friend, Paul Langevin, praised his skills as a mentor. He wrote that "my finest memories of my school years are those of moments passed there standing before the blackboard where he took pleasure in talking with us, in awakening in us fruitful ideas, and in discussion of research which formed our taste for the things of science." Henri Poincaré recalled the night before his death when he sat next to him "and he talked with me of his plans and his ideas." He continued by lamenting the "stupid accident" that took the man through whom he was better able to understand the "grandeur of human intelligence."[9]

The house in Paris seemed to Marie to be haunted by memories of Pierre. The family, including Pierre's father, moved to a house in Sceaux near where Pierre had grown up. She had several reasons for moving there, although she referred to her new home as a "house without charm."[10] This charmless house had a place for old Dr. Curie to cultivate a garden and an area where the girls could have a playground with a crossbar, a trapeze, rings, and a rope.

EDUCATION OF THE CHILDREN

Marie felt that physical exercise was an important part of a child's education. She insisted that Irène and Eve participate in sports such as gymnastics, swimming, bicycling, horseback riding, rowing, skiing, and skating. Interestingly, given her propensity to overwork, she wanted to give her girls respite from their books and time to play in the open air. In order to assure that Irène (Eve was too young) got the kind of education that she approved of—one with a heavy emphasis on the sciences—she developed a home schooling experiment. She organized a cooperative school with approximately ten children attending and co-opted the parents of the pupils and willing friends to be teachers. Henriette Perrin taught history and French; Alice Chavannes taught English, German, and geography; Henri Mouton of the Pasteur Institute taught natural science; and Paul Langevin was talked into teaching mathematics. Jean Perrin taught physics and Marie Curie taught chemistry using their own laboratories. Instead of class, the little group sometimes visited museums in Paris. The home schooling experiment lasted only two years, for their parent teachers were too overworked to continue the project. An additional factor hastened the demise of the experimental school. Because the students would eventually have to take a baccalaureate examination, it was important that they become involved in an official program.

Much later, Marie attempted to get Irène into an all-male school, the Lycaeum Lakanal. The *New York Times* of April 9, 1911, reported the incident under a headline, "Mme. Curie Will Persist." The article explained that "the co-discoverer of radium, has declared war on the old-fashioned French prejudice against mixed schools." This created a furor among the professors of the institution. She asked the president of the lyceum to admit the then sixteen-year-old Irène to the regular course of study followed by young men. After his hasty refusal, she did not give up and brought it up before his superiors. Convinced that the men received a better education, she planned to bring the question before the minister of education. The request was not granted. The article suggested that it was not the parents of the male students who objected, but the professors. In an interview, one stated:

> The teaching of girls is the most dreadful ordeal for male professors. In addition to being criticised for one's appearance, the slightest negligence in dress, or the shortest hesitation as to a date, or such like, make a teacher liable to be held up to ridicule and lose all his grip of discipline on the whole class.

He concluded by stating that he would rather lecture before one hundred boys than twenty girls.[11]

Grandfather Curie kept the house from being a humorless, silent place. Marie decided that she would protect the girls from sorrow by never mentioning their father's name. Eve clearly felt that this was a mistake, writing that "rather than plunge them [Irène and Eve] into an atmosphere of tragedy, she deprived them, and deprived herself, of noble emotions."[12] She was morose yet demanding of the girls. Eugène Curie, on the other hand, was joyful, teasing, and full of fun. His blue eyes sparkled as he played with the girls. He also made natural history and botany enjoyable for Irène (Eve was still too young) by combining instruction with fun. In 1909, an illness confined him to his bed for a year. He was a difficult patient and Marie spent much of her time pacifying and distracting him from his illness. He died on February 25, 1910. Marie requested that the gravediggers remove Pierre's coffin and place Eugène's at the bottom of the grave with Pierre's coffin on the top. Even in death she wanted Pierre near her.

Although the girls achieved a good academic education, their somber, silent mother affected their social skills. Irène especially would hide when strangers would visit, and she would duck her head when spoken to. Eve tells of the time that Marie punished Irène for impudence by not speaking to her for two days. In spite of the sad household, Irène and Eve loved their mother as dearly as she did them. They spoke of her as "Darling Mé or Sweet Mé." Although Marie was too reserved to allow her grief to be seen, her sad eyes and nervous habit of rubbing her radium-irritated fingers together made the girls sympathetic. Eve showed some resentment when she described her childhood, explaining that "in spite of the help my mother tried to give me, my young years were not happy ones."[13] Irène and Eve had very different personalities. Although Eve was prettier and more approachable, Irène was much more like her mother. Irène's interests also paralleled those of Marie, but Eve loved to write and was musically inclined.

NOTES

1. Eve Curie, *Madame Curie: A Biography* (Garden City, NY: Doubleday, Doran & Co., 1938), p. 246.
2. Ibid.
3. "Prof. Curie Killed in a Paris Street," *New York Times,* April 20, 1906, p. 11, col. 3.
4. Susan Quinn, *Marie Curie: A Life* (New York: Simon and Schuster, 1995), p. 232.
5. Eve Curie, *Madame Curie*, p. 249.
6. Ibid., p. 252.
7. "Pension for Curie Family," *New York Times*, April 22, 1906, p. 9, col. 2.
8. Eve Curie, *Madame Curie*, p. 259.
9. Marie Curie, *Pierre Curie*, trans. Charlotte and Vernon Kellogg (New York: Macmillan, 1923), pp. 146–51.
10. Quinn, *Marie Curie: A Life*, p. 248.

11. "Mme. Curie Will Persist," *New York Times*, April 9, 1911, pt. 3, p. 1, col. 6.

12. Eve Curie, *Madame Curie*, p. 268.

13. Ibid., p. 272.

Chapter 8

SCANDAL!

Marie soon regretted her generosity in giving away her personal radium. She positively impressed American Andrew Carnegie, the wealthy philanthropist and author of *The Gospel of Wealth,* whom she met in Paris shortly after Pierre's death, and he decided to endow her research. She was grateful, because it allowed her to finance a research staff around which she could build a school of radioactivity in Paris. The endowment suited her very well, because she could accept the money for her students, not for herself.

By 1906 almost everybody accepted radium as a new element. But there was one important exception. The grand old man of English science, Lord Kelvin, had never embraced it. On August 9, 1906, he chose to present his ideas in the famous *Times* of London. He selected this popular venue because radium and everything about it fascinated the public. He wrote that radium instead of being a new element was only a compound of lead with five helium atoms. Of course, if he was correct, then both Marie Curie's work and Rutherford and Soddy's theory of radioactive disintegration would be in tatters. Kelvin, at eighty-two, just could not conceive of the possibility of a new element. And of course Marie was equally certain that she was correct. Kelvin had called her basic assumption that radioactivity was an atomic property into question. Even though she, and most of the scientific community, accepted radium as an element, she felt challenged to produce a purer form than the radium chloride that she had purified earlier. Thus, she began the purification process again in her new laboratory. By 1907, she had produced a perfectly pure radium chloride, allowing her to determine an even more precise atomic weight for the element radium.[1] By the time she had completed this work, there could be little doubt that radium was a new element.

Nevertheless, Lord Kelvin still clung to his ideas tenaciously. The idea of one element being transformed into another was unacceptable to Kelvin and others who believed that it smacked of alchemy. At least three other distinguished scientists had conducted experiments that indicated transformation occurred in other elements: William Ramsay (1852–1916), Ernest Rutherford, and Frederick Soddy (1877–1956). For her part, Curie was hesitant to give her opinion of Sir William Ramsay's reported transmutation of copper into helium. However, she finally relented and on August 18, 1906, the Sunday *New York Times* quoted Curie as saying that she shared the opinions of Ramsay, Rutherford, and Soddy and would place radium in a group of unstable elements. Radium, she hypothesized, is composed of atoms that undergo spontaneous transformation resulting in helium as one of its products. But she still was not absolutely convinced that this transformation was the source of the helium. She also considered the possibility that it was in the gases surrounding the radium, which were never completely removed even in a vacuum. In either case, there was no doubt in her mind that an atomic transformation had occurred. Marie did not resent Lord Kelvin's ideas or those of any scientist whose views differed from hers. Only from a free and open discussion of ideas, she reported, can additions to knowledge be made. Kelvin was an old man and set in his ways by this time, and Marie may have tried to avoid offending him. She wrote that she did not consider combating Lord Kelvin's opinion useful. She did reiterate that radium was a distinct chemical element. Lord Kelvin died in December 1907 and the opposition to the new element evaporated. Marie, with the assistance of André Debierne, also confirmed that polonium was a new element.

In 1908, the *New York Times* proclaimed that there was practically no commercial use for radium, but that its value for laboratory experiments created a demand that could not be satisfied because of the great cost. A factory in France supplied radium bromide, not pure metallic radium, at a cost of $40 million a pound. The reporter admitted that he was still confused as to the nature of radium—whether it was a substance or a quality. "To speak of a pound of it is like speaking of a pound of sunlight."[2] After reporting that fresh radium experiments were being performed daily in Marie Curie's laboratory at the Sorbonne, the reporter concluded that the secret of radium was still unsolved. Of course, articles such as this one and those scientists who doubted that radium was an element inspired Marie to double her efforts to obtain pure metallic radium.

A short note on the front page of the Tuesday, September 6, 1910, edition of the *New York Times* proclaimed that Curie "had announced today to the Academy of Sciences that she had succeeded in obtaining pure radium." The article continued by adding that previously it had existed only in the form of salts. In order to obtain the pure radium, Curie treated a decigram of bromide of radium by an electrolytic process, "obtaining an amalgam which was extracted from the metallic radium by distillation." According to the report, the radium "has the appearance of a white metal, and is capable of adhering strongly to iron." The white metal changes to black when exposed to the air, "burns paper, and oxidizes in water."[3]

The next major project that Marie and her institute became involved in was the development of an international standard for radium. It was essential for a researcher who was working with radium to know its purity. Since hospitals were using radium in the treatment of cancer, they too needed to know its purity so that they could determine an optimum dose to treat the tumors. Although national rivalries came into play in determining the international radium standard, all agreed that Madame Curie's eminence gave her the right to prepare the standard. In 1911 she established the standard and deposited a thin glass tube a few centimeters long with the pure salt inside at the International Bureau of Weights and Measures near Paris.

Marie Curie had managed to make a number of enemies along the way. At international conferences she appeared to be uncompromising, determined, and demanding. Although the young people in her laboratory adored her, those who did not know her well felt rebuffed. They did not realize that her apparent coldness and refusal to engage in light conversation was due in large part to her shyness and sensitivity.

Some of the enemies that she made returned to haunt her later. One of the most vicious was an American, Bertram Borden Boltwood, who often suffered from bouts of depression. Rutherford admired his skill as a radiochemist and enjoyed his company when he was feeling cheerful and full of fun, and they became fast friends. Unlike when Pierre was alive, Marie was more proprietary with the discoveries from her laboratory. When Boltwood asked Marie Curie to allow him to compare one of his radium solutions with her own radium standard she refused to do so. In a letter to Rutherford he wrote, "The Madame was not at all desirous of having such a comparison carried out, the reason, I suspect, being her constitutional unwillingness to do anything that might directly or indirectly assist any worker in radioactivity outside her own laboratory."[4] The diplomatic Rutherford was able to get Marie to lend him the standard when he needed it several months later.

Although Rutherford liked Curie as a person, he began to have less respect for her originality. Both he and Boltwood attributed her success more to hard work and tenacity than to any innate creativeness. After she published her comprehensive summary of advances in radiation, *Treatise on Radioactivity* (1910), Rutherford, although publicly supportive (he had reviewed the book for the journal *Nature*), privately was condescending. Her health was clearly declining, and she often missed conferences and meetings because of illness. Some of her more unsympathetic colleagues thought that she was malingering when a topic to be discussed was not to her liking. One example had to do with naming the unit of measurement for the Radium Standard the "curie." There was disagreement over what this unit should be. When the decision did not go her way, she left the meeting claiming illness. She wrote a note on the hotel notepaper that if the name of Curie was to be adopted then it was she who would define it. She won the battle but also made more enemies. Pleading a bad cold, she did not attend the congress's festive banquet.

In spite of his general support for Curie, Rutherford found Marie a difficult person with whom to work. In the spring of 1910, he approached her with a problem of different radium standards—he had his own, Marie had her own, there was one from Vienna, and, no doubt, there were others in different parts of the world. They saw the need to appoint a committee to develop an international standard. Without such a standard there would be no way to check the agreement between results obtained in different laboratories. Since medical applications of radium were increasing, the creation of a universal standard was vital. Marie agreed to prepare the standard, but she informed Rutherford at the first Solvay Conference that she wanted to keep it in her own laboratory partly because of sentimental reasons and partly because she wanted to make additional observations on it. This attitude was different from her early perspective, when she and Pierre agreed that radium belonged to the world—not to one country or to any individual. When Rutherford explained that the International Committee could not allow the standard to be in the hands of a single person, Marie was distressed. By being diplomatic he finally was able to negotiate a solution, but there were many other questions where he felt that she was being unnecessarily difficult. The solution involved comparing the Viennese standard of Stefan Meyer with Marie's standard. Although Rutherford assumed that the two standards would closely agree, he knew that if they did not, Marie would be more than agitated.

The two standards were compared in March 1912. The accommodation may have been made easier because of Marie's illness, for she was not well enough to accompany the others to her laboratory. To everybody's relief, they found that the two samples agreed.

PAUL LANGEVIN

One day, five years after Pierre's death, Marguerite Borel reported that Marie appeared in the Perrins' dining room in a white dress with a rose at her waist. Marguerite, much younger than her husband, Èmile, was the type of person who invited confidences and was not immune to gossiping about what she heard. She claimed that when Marie replaced her somber dark clothes with the sparkling white dress, she knew that Marie was no longer mourning Pierre. Who had caused this abrupt change of heart? One of Marguerite's confidantes was Pierre's former student, Paul Langevin. He had previously poured out the details of his unhappy marriage to her receptive ears. Marie had also confided in her, excoriating Langevin's wife for being unsupportive of the brilliant scientist that Paul was. Marguerite suspected that a romance was blossoming between the two.

Several events and inventions made it difficult to keep the romance hidden. Three inventions, the linotype, the electric telegraph, and the telephone, made possible a larger newspaper with headlines and pictures. The improved newspapers carried news stories designed to appeal to the lower socioeconomic classes as well as the elite. Several of these newspapers had various axes to grind. For example *L'Action Française* first appeared in March 1908, and was basically an anti-Jewish scandal sheet. Another new development was the beginning of science journalism—writing about science for a popular audience. An example of this new form was the periodical *Figaro*.

Not only a press hungry for scandal, but an event that occurred publicized by that same press, brought Marie Curie to the attention of the newspaper-reading public. Marie made a decision to become a candidate for the Academy of Sciences (Académie des Sciences). Her determination to stand for election seemed strange in light of Pierre's bad experience with this very traditional organization. But, if she were elected, she could expect more money to come to her laboratory as well as enjoy enhanced prestige both to herself and to her laboratory. Unlike Pierre, these things were important to her. A woman had never been elected to the academy and it seemed that Marie could not resist the opportunity to be the first. Thus, she swallowed her pride and made the required visits to the academy members.

Her decision to become a candidate immediately became front-page news. At first the articles were full of praise. *Figaro* published a long article extolling her virtues as the grieving widow who managed to hide her personal sorrow and continue to work as one of France's most respected scientists. But Marie had also managed to collect a group of influential enemies. Her outwardly cold, superior attitude annoyed those very scientists whom she needed for support. Thus, while they might have accepted the virtues of a modest, feminine woman, her opinionated stance fostered jealousy in these colleagues. Others such as Rutherford and Boltwood maintained their doubts about her scientific creativity.

The entire idea of women being admitted into the traditional male bastion of the academy became a media circus. One respectable newspaper, *Le Temps*, was sympathetic but not completely supportive of Curie. However, in the France of this time, there existed a strident, radical right-wing press. One of the most shrill of these voices came from *L'Intransigeant*. There was none of the reasoned opposition that characterized the more moderate press. They accused Marie of being little more than a hack. They claimed that all of the important discoveries were made during Pierre's lifetime. The attacks became personal. What did this woman think she was doing putting herself forward for membership in this male institution? In the United States, the *New York Times* reported on the controversy. The article described a lively two-hour-long discussion in the academy about the admission of women. The committee was divided into two camps, one favorable to Curie and the second opposed. The opposition considered that the admission of women would be an "audacious precedent."[5]

There was only one thing missing in the scenario—a credible opposition candidate. One appeared in the form of sixty-six-year-old Edouard Branley (1844–1940). Branley, like Curie, was retiring and undemonstrative. Two times previously he had been a candidate for the academy and two times before he had been defeated. Many French people felt that his contribution to modern technology (he made what he called radio conductors from tubes of iron filings that could receive electromagnetic signals) should have earned him a Nobel Prize in Physics along with Guglielmo Marconi (1874–1937). The right-wing press, especially *L'Intransigeant* and *L'Action Française*, was delighted to find a suitable male candidate. *Figaro* cheerfully leapt into the fray, proclaiming that the upcoming election would represent the "battle of the sexes." The conflict got especially nasty with the liberals, feminists, and anticlerics on the side supporting Curie and the nationalist, pro-Catholic, anti-Semitic right wing supporting Branley. The right-wing press sneeringly brought up Curie's Polish origins and this anti-Semitic press even implied that in spite of her Catholic parents she had Jewish origins.

The actual election came on January 23, 1911, and there were accusations of cheating on both sides. When the vote was finally announced Curie had twenty-eight votes, Branley twenty-nine, and a third candidate had one. To get a clear majority a second vote was necessary. The result was that Curie remained at twenty-eight and Branley got thirty. Marie's disappointment was intense. She never again stood for membership in the academy.

The controversy had an amusing by-product. An inventor who falsely portrayed himself as the Comte de Chambert was charged with swindling a wealthy old woman by convincing her to invest in a system for restoring ancient paintings by the application of electricity. The inventor was vague about the nature of his product and he was able to garner some support among some important men. The defense stated that it would call M. Branley, Marie's successful opponent in the academy election, and the prosecution responded by threatening to call upon Marie Curie as an expert witness.[6]

By 1911 the name Marie Curie was a household word. She had been the subject of both positive and negative publicity through her search for membership in the academy. Her work with radium was exciting and romantic, and the fact that it was a woman working in this field made it even more so. But Marie attempted to remain a private person. As far as anyone knew, she stayed a grieving widow without a romantic interest. This seemed to be the case until one day in the spring of 1911 when Jean Perrin and André Debierne explained to Marguerite Borel that a group of letters from Marie Curie to Paul Langevin had been stolen. The thief had broken into Paul Langevin's study. The letters implied a close relationship between the two scientists and if the newspapers got hold of them it would mean a disaster for Marie's reputation.

In late nineteenth- and early twentieth-century France many men kept mistresses. It was acceptable practice as long as a man appeared with his wife at social functions and observed the prescribed niceties. If they were both discreet, he was free to support a mistress. But if the affairs became public, they were universally censured. This view would have important repercussions on Marie Curie as a famous woman. A mistress of humble origins would not have raised the public's ire, but Marie Curie was a career woman who supported herself and who was known by most French people. Newspaper reports on her unsuccessful candidacy for the French academy suggested that the public was ready to denounce her. *L'Intransigeant* stated that she should have withdrawn from the race as homage to Branley, an older man, by a woman. This newspaper complained that the gesture was not made.

Madame Langevin and the newspapers were quiet for eight months after the letters were stolen. Perhaps she was still hoping that the marriage could be repaired. However, the relationship between Langevin and his wife, Jeanne, deteriorated to such a degree that after an argument he snatched their two boys and left the house. Jeanne claimed that Paul had struck her in the face for cooking badly. He claimed that she had hurled insults at him in front of the boys. At any rate, the fact that he had taken the children without his wife knowing where they were made him a candidate for a lawsuit. There is a possibility that Paul was paying blackmail to keep the letters under wrap. During this time, Marie loaned him a considerable amount of money.

Marie was not oblivious to the damage to her reputation and the effect on her daughters if the letters were made public. In a letter to Langevin quoted by biographer Susan Quinn she speculated on what would happen if Jeanne were to have another child. She concluded that they would be "judged very severely." She also told him that "I can risk my life and my position for you, but I couldn't accept this dishonor in the face of myself, of you and of people I esteem."[7]

While Marie was in Brussels attending a radiation congress, the Solvay Conference, the vague rumors burst into a full-blown scandal. One of Paris's most famous newspapers, *Le Journal,* launched a front-page story under the headline "A Story of Love: Madame Curie and Professor Langevin." Perhaps Jeanne Langevin was especially jealous because both Paul and Marie attended this conference. At any rate she went to the paper with the letters. Jeanne's mother also supplied material for the article, all of which was damning to both Marie and Paul. In the article, Jeanne came across as a mistreated wife who had only gone to the newspapers for the sake of her children. Marie was portrayed as a woman who was engaged in the masculine pursuit of science—a harridan who specialized in taking another woman's husband and spoiling the lives of their children. However, there were doubts about the truth of the story. Supporters claimed that many of the implications drawn from the letters were merely the rantings of a jealous wife. And when Marie returned from Paris, she wrote a scathing denial. She sent a letter to *Le Temps,* excoriating the press for its intrusion into her private life. She averred that there was nothing in her actions of which she was ashamed. She also threatened to demand monetary damages to be used in the interest of science. The *Journal* reporter retracted the story. In fact, he was abjectly apologetic. Both Marie's letter and the reporter's denial were widely publicized, and it seemed as if the accusations would die away.

Perhaps the charges would have evaporated if the public and the newspapers had not been reminded of it when Curie was nominated in

1911 for her second Nobel Prize, this time in chemistry. The Nobel committee stressed the importance of Curie's work in obtaining radium in its pure metallic state. They also stressed the medical use of radium in treating cancer. The importance of radium led the committee to urge the Swedish academy to award a second Nobel Prize to the same person. However, the very staid academy was quite concerned after the newspaper broke the story of the scandal. It was mollified after the denials by Marie and by the man who had written the article. On November 7, 1911, the academy voted to award her the prize. The newspaper publicity was much different from that which occurred when she was selected for the first prize with Pierre and Becquerel. The newspapers basically ignored her. Although *Le Temps* produced a front-page article on the Nobel Prize, it was totally devoted to the prize winner in literature, Maurice Maeterlinck. The recipient in chemistry was not even mentioned. Her scientific friends, including Albert Einstein, were very supportive. Jacques Curie implied in a letter to Marie that he would support her even if letters came out indicating that she and Langevin had had an affair.

Jeanne Langevin planned a gigantic blackmail scheme. If Langevin did not give up custody of the children and pay her one thousand francs a month, she threatened to give the letters to the newspapers. When Langevin did not accept the terms, Jeanne charged her husband with "consorting with a concubine," an accusation that would be heard in a criminal court. Marie's lawyer insisted that the results of a trial would be favorable to her, and that she could go to Sweden to accept the Nobel Prize. Friends of both Curie and Paul Langevin attempted to cover up the evidence. The newspaper *L'Action Française* was not discouraged by the impressive array of people supporting Marie. Instead they used the occasion not only to blast her as a home wrecker but also to insert racist remarks about her as a foreigner. Another newspaper, *L'Intransigeant,* jumped into the fray and castigated Marie and Paul while insisting that Jeanne should definitely have custody of the children. Neither of these two papers actually published the letters.

On November 23, 1911, a vitriolic weekly, *L'Oeuvre,* published excerpts from the letters Marie Curie and Paul Langevin had written to each other. Although nothing explicit was included, it seemed obvious to the readers that there was truth to the accusations that they were having an affair. One letter from Marie was especially incriminating. In this letter she outlined the steps by which Langevin could remove himself from his marriage. The French public was incensed by these letters. The magazine pitted the foreign woman, Curie, against a respectable French woman, Jeanne Langevin. The right-wing press became even more raucous.

The result of the hate campaign was frightening. Hostile crowds gathered around the Curie house shouting imprecations. She and the children were rescued by Marguerite Borel and André Debierne who spirited Marie and Eve off to the Borel apartment. Little Eve had no idea of what was occurring as she clutched her mother's hand. Fourteen-year-old Irène, on the other hand, was very aware of what was happening. When Debierne collected her from her school (gymnasium), she had already read the scathing article in *L'Oeuvre*. When she saw her mother, Irène clung to her and both mother and daughter seemed completely numb. Finally, Henriette Perrin was able to take her to their home. Eve was looked after by a maid, and Marie was able to lie down quietly.

Peace eluded Marie for a long time. The public gleefully took sides in the conflict. Marie's supporters took the view that she was innocent of the charges and was being persecuted by her enemies. Her detractors insisted that she had defiled French motherhood.

Although Marie was unaware of it, a duel was being fought by her supporter, Henri Chervet of *Gil Blas*, and her detractor, Léon Daudet of *L'Action Française*. Although Daudet was the more experienced dueler, he suffered a deep wound in his elbow. Newspapers in the United States picked up on this event. The *New York Times* described the duel as a dispute over the merits of the charges, which Mme. Langevin instituted against her husband. The duel was fought with swords. The tone of the article was sympathetic to Marie, explaining that the allegations were based on the fact that Langevin and Curie worked together. This proximity caused "a jealous feeling on the part of Mme. Langevin, who thereupon brought suit against her husband, coupling with his name that of Mme. Curie."[8]

This duel was just the first of five to be fought, provoked by the Langevin/Curie affair. A second duel between Pierre Mortier, a writer for *Gil Blas* and a supporter of Curie, and Gustave Téry of *L'Oeuvre* extended the farce, according to a *New York Times* article of November 25, 1911. Although Téry had apologized for writing an article that Curie and Langevin had eloped, he had obtained an illegal copy of the complaint that Mme. Langevin had filed against Curie and Langevin. The duel resulted in Mortier being wounded in the arm.

The most famous of the duels was that between Langevin and Téry. Since Téry had described Langevin as a "cad and a scoundrel," Langevin felt that his honor would be impugned if he did not challenge Téry to a duel. Langevin had a difficult time in finding seconds, because his academic friends, although sympathetic, were not interested in becoming involved in the conflict. Langevin finally located two friends who reluctantly agreed to be his seconds. They met on the morning of November 26 and chose pistols as their preferred weapons. The tall, thin Langevin arrived first. Gustave Téry then appeared with his entourage. The morning was gray and foggy. Each second gave a pistol to his person. The second, who was chosen by lot to direct the proceedings, asked the combatants if they were ready. After an affirmative answer he counted, "one, two, three, fire!" Langevin raised his pistol arm up as if to fire. Téry, however, kept his pistol barrel pointed toward the ground. When Langevin saw that Téry did not intend to fire, he lowered his gun. Téry had felt that he could support Jeanne Langevin by killing Paul. However, he had a change of heart when he realized that by killing Langevin he would be depriving France of one of its most famous scientific minds. The newspapers reported the duel in great detail.

It is unclear as to when Marie Curie was informed of Paul Langevin and Téry's duel. Since it was the talk of Paris and even the Nobel Committee in Sweden knew about the duel, it is improbable that she was shielded from it. Marie had asked the scientist Svante Arrhenius (1859–1927), her enthusiastic supporter for the Nobel Prize and an important member on the Nobel Committee, if she should go to Sweden to accept the prize because it was likely that the press was going to stir up ugly rumors. He first assured her that she should go to Sweden where she would be considered an honored guest of the nation; however, after the Langevin duel and the publication of the letters, he reneged. He said that the duel gave the perhaps false impression that the published correspondence was true. He and his colleagues agreed that Marie should stay in France. If the academy had believed that the charges were authentic, continued Arrhenius, it probably would not have given her the prize in the first place. Disappointed by Arrhenius's response, she wrote him rather truculently that she saw no connection between her scientific work and her private life. She stated that she could not accept the idea that the appreciation of her scientific work was influenced by libel and slander concerning her private life.

Marie was somewhat naive in thinking that as the most important woman scientist in the world, people would be uninterested in her private life! Both she and her lawyer wanted the case to go to trial, but Langevin decided to admit that he was in the wrong. Loath to take sides against his wife in public, he feared that their four children would suffer. There was a cash settlement and Jeanne Langevin got custody of the children. Paul was allowed some visitation rights and was able to direct their intellectual development. When they were in their mid-teens, the boys would come to live with their father. Many years later, Marie Curie's granddaughter and Paul Langevin's grandson married without having any clue about the scandal.

The controversy was not limited to France, for in addition to reporting on the duels, the *New York Times* presented editorials, letters to the editor, and numerous additional articles. In order to see how the presentation of the subject was different in the United States and France, it is interesting to look at some of the articles. An editorial on November 24, 1911, the day after *L'Oeuvre* published excerpts of the letters, in the *New York Times* presented a different interpretation of the subject. The editorial stated that the honor of science and that of the Sorbonne could never be tarnished because an extraordinary Frenchwoman "has been made, rather late in life, the heroine of a somewhat scandalous romance." Clearly this newspaper blamed Jeanne Langevin, writing that Marie had been "assailed by a jealous woman who accuses her of estranging a husband from wife and children." Without judging whether or not the charges were justified, the editorial concluded that people of extraordinary abilities sometimes defied social conventions. The article concluded that "there are hints of deliberate mischief-making in the case. The letters quoted by Mme. Langevin's complaint may not be genuine. In any case neither science nor the Sorbonne can suffer at all from social scandals affecting the lives of scientists."[9] In response, a letter to the editor agreed that gifted people may legitimately have different standards than ordinary people. The writer also quotes a statement: "Don't believe anything you hear and only half of what you see."[10] Another article suggested that even if the accusations proved true, "Mme. Curie will have to pay the cruel penalty of her sex and be known hereafter much less as the co-discoverer of radium and one of the most eminent laboratory workers that has ever existed than as the woman who stole another woman's husband."[11] By December 4, 1911, the *New York Times* printed a number of rumors that had been spread about the affair. One sensational one stated that Mme. Langevin "proposed to allow her husband six months' absolute liberty, so that he may experiment with Mme. Curie's companionship as much as possible before coming to a decision." Jeanne Langevin's sympathizers hinted that this was a maneuver by Marie's supporters to test public opinion. An even more unsubstantiated tidbit of gossip in scientific circles suggested that Professor Langevin had helped to assure Curie's scientific reputation. Although the rumors were reported in the United States, the general tone was that they were ridiculous.[12] On December 14, the *New York Times* reported that Mme. Langevin had filed for divorce. She "asks in her petition for separation from her husband and that the children of the marriage shall be left in her custody. She also claims separate maintenance."[13]

On December 23, 1911, the *New York Times* reported that the Langevin case had been dismissed. The criminal charge against Langevin was withdrawn and his wife reported that she was "completely satisfied with the decision of the Divorce Court granting her a separation from her husband and the custody of her four children."[14] Mme. Curie's name was not brought up in the proceedings. On December 17, 1911, the Sunday *New York Times* summarized the events surrounding the scandal.

Marie endured the scandal stoically at first, holding her head high and pretending to be unphased by the accusations. However, it became apparent as the scandal crossed the ocean and entranced the American public nearly as much as the French and appeared in other publications throughout the world, that Marie's apparent assurance began to fade. She began to dread her encounters with people. The fact that she was the most outstanding woman scientist in the world, one who at this very time was to go to Sweden to present a lecture for a second Nobel Prize, made her indiscretion all the more interesting to the public. What might have been one of the most satisfying times in her life had turned to a nightmare resulting in ostracism and illness.

NOTES

1. Marie Curie, "Sur le poids atomique du radium," *Compte rendus* 145 (1907): 422.
2. "Practically No Commercial Use," *New York Times,* May 3, 1908, pt. 5, p. 10.
3. "Pure Radium Obtained," *New York Times*, September 6, 1910, p. 1.
4. Bertram Boltwood to Ernest Rutherford, October 11, 1908, Cambridge University Library, quoted in Robert Reid, *Marie Curie* (London: Collins, 1974), p. 167.
5. "Dispute over Mme. Curie," *New York Times,* December 4, 1910, p. 4, col. 3.
6. "Mme. Curie May Testify," *New York Times,* February 5 ,1910, pt. 3, p. 4, col. 2.
7. Marie Curie to Paul Langevin, summer of 1910, reproduced in *L'Oeuvre,* in Susan Quinn, *Marie Curie: A Life* (New York: Simon and Schuster, 1995), p. 296.
8. "Editors in Duel Over Mme. Curie," *New York Times,* November 24, 1911, p. 3, col. 4.
9. "Madame Curie," *New York Times,* November 24, 1911, p. 12, col. 2.
10. "To the Editor," *New York Times,* November 26, 1911, p. 14, col. 5.
11. "A Scandal and Its Consequences," *New York Times,* November 29, 1911, p. 10, col. 5.
12. "Curie Case Rumors Are Not Verified," *New York Times,* December 4, 1911, p. 4, col. 3.
13. "Langevin Divorce Suit," *New York Times,* December 14, 1911, p. 9, col. 2.
14. "Langevin Case Withdrawn," *New York Times,* December 23, 1911, p. 3, col. 5.

Chapter 9

THE SECOND NOBEL PRIZE, ITS AFTERMATH, AND WAR

In the midst of the furor surrounding the scandal, Marie, Bronia, and Irène went to Stockholm to attend the Nobel ceremonies. She gave her acceptance speech with dignity, and from her demeanor no one would have realized her inner turmoil. Standing in front of her scientific peers who clearly had reservations about her receiving the prize, Marie began:

> Some fifteen years ago the radiation of uranium was discovered by Henri Becquerel, and two years later the study of this phenomenon was extended to other substances, first by me, and then by Pierre Curie and myself. This study rapidly led us to the discovery of new elements, the radiation of which, while being analogous with that of uranium, was far more intense. All the elements emitting such radiation I have termed *radioactive,* and the new property of matter revealed in this emission has thus received the name *radioactivity.*[1]

Curie continued by stating that the task of isolating radium (for which she received the prize) "is the corner-stone of the edifice of the science of radioactivity." Noting that because radium "is the most useful and powerful tool in radioactivity laboratories" the "Swedish Academy of Sciences has done me the very great honour of awarding me this year's Nobel Prize for Chemistry."[2] In this speech, Curie used the first person more than usual, noting that I did this or I did that, although she gave extensive credit to others who had done work on radioactivity. To all outward appearances the recipients of her lecture were accepting and the scandal did not seem to play a part in the way the conservative scientists listened to her speech.

Physically, however, her health broke down, and she was rushed to a hospital eighteen days after her Nobel lecture. She was gravely ill with a severe kidney ailment. Marie's close friends thought that her collapse was precipitated by the fallout from the Langevin affair. However, her doctors diagnosed an infection in the kidney and ureter caused by some old lesions. Although they recommended surgery, the doctors preferred to wait to see if the infection would go away without the more radical treatment. During January 1912, she was cared for by the Sisters of the Family of Saint Mary. The acute disease lessened and she returned home and was reunited with Irène and Eve. Although she still needed kidney surgery, she went back to work in her laboratory in early March in a weakened state. She was especially frail when the time came for the operation in late March. Although the surgery was a success, her health was compromised for many months.

The next months were a nightmare for Marie. Physically, she was still weak, but her state of mind was more important. She sank into a deep depression, and her friends feared that she might take her own life. In March she returned to the hospital for surgery to remove the lesions. Afterward she was so ill that she believed that death was impending. She even made plans for the disposition of her affairs, including her radium. The recovery from the surgery took much longer than was expected. In 1909 she had weighed about 123 pounds but after her operation she weighed only 103 pounds. She moved from one convalescent center to another, often leaving her daughters behind. She was subject to painful spasms that kept her away from teaching much longer than she had expected. But it was not just her physical pain that haunted her.

Feeling that she had disgraced the name Curie, she kept her address secret from everyone but her family and a few close friends who were caring for the children. The habit of recording every mundane detail of her life remained with her even in this dark period of her life. She still noted the price for laundry, drugs, and lessons and clothes for the children. In spite of the hateful letters and malicious articles that appeared in the press, many strangers and friends rallied around Marie. Her brother, Jòzef, and her sisters, Bronia and Hela, rushed to France to give her their support. Perhaps her most ardent defender was Jacques Curie, who might have been expected to react adversely. Bronia rented a small house for her outside of Paris under the name of Dluska. Her physical problems were not over. In June, she had a relapse and was taken to a sanatorium in the mountains of Savoy.

Marie's illness and disgrace was especially difficult for Irène. The mixture of child/young woman that Irène was at this time made the suffering of her beloved mother extremely hard to endure. Irène was deeply hurt when her mother told her that she was not to use the name Curie when addressing her letters, but Madame Sklodowska.

During her visit to the Royal Institution in London in 1903, Marie had met a second collaborative couple (in addition to Margaret and William Huggins), the English woman scientist Hertha Ayrton (1854–1923) and her husband. The two women became good friends. Although Marie's and Hertha's scientific accomplishments had little in common, there were similarities in other aspects of their lives. For instance, both women were married to scientists with whom they collaborated. Ayrton was married to W. E. Ayrton (1847–1908), a fellow of the Royal Society, a pioneer in electrical engineering, an avid advocate for technical education, and a zealous supporter of women's rights. Hertha entered science through invention. Based on an idea of her cousin Ansel Lee, she invented an apparatus fashioned to divide a line into any number of equal parts, an instrument she claimed was useful to artists, decorators, engineers, and ship's navigators. The backgrounds of the two women were quite different, although each had encountered adversity in her younger years. Hertha, born Phoebe Sarah Marks, was born in Portsea, England. Her father, Levi Marks, was a Polish-Jewish refugee who had trouble making ends meet in his clock making/jewelry trade. Levi died in 1861 and Hertha's mother tried to support the family by her needlework. Hertha helped care for her siblings and learned to sew, cook, and keep house so that her mother could spend all of her time on needlework. The only way that she was able to get an education was through an aunt who ran a school in London. Marie Curie and Hertha Ayrton appeared to be quite different in their personalities. Whereas Marie was shy and retiring, Ayrton was self-assured and appeared to many to be abrasive. As an expression of independence, Ayrton rejected her given name and adopted a new one, Hertha, suggested by her friend Ottilie Blind.

Both women, however, were stubborn and dedicated to science.

Both had shunned the religion in which they were raised. For Marie, this religion was Catholicism; for Hertha, Judaism. Marie was a devout Catholic in her younger days before the deaths of her mother and sister. Their deaths led her to question why God would have allowed these tragedies to happen. Like Marie, Hertha, a devout Jew in her younger days, became a skeptic after her association with her atheistic cousin Marcus Hertog.

Because of financial problems both women found it difficult to obtain an education. Hertha found her educational savior in the form of Barbara Bodichon, an eccentric philanthropist who was interested in women's causes and who was one of the founders of the women's college Girton College, Cambridge. Both Marie and Hertha worked as governesses to save money for their university educations. After working for six years as a governess, Hertha explored ways to continue her education at the newly established Girton College. She was introduced to Bodichon, who suggested that she take the scholarship examinations. However, she failed to win either of the two openings. Nevertheless, Barbara Bodichon with the help of other friends managed to scrape together enough money to allow her to enter Girton.

Hertha met her physicist husband, Professor Ayrton, during her brief teaching career after she left Girton. Unlike in the case of Pierre and Marie, W. E. Ayrton was much older than Hertha. He was a widower with one daughter, Edith, and believed in equality of opportunity between men and women. The strongly independent personalities of both Ayrtons made their relationship more of a mutually supportive one than an actual partnership such as characterized the Curies' association. But the women in the case of both couples found succeeding in science much easier because of their scientist husbands. Some of their experiences as scientists also were similar. Hertha Ayrton had been denied membership in the Royal Society as had Marie Curie in the Académie des Sciences. In addition to W. E. Ayrton's daughter, Edith, the Ayrtons had one daughter together, Barbara Bodichon Ayrton.

Just as Marie Curie had been passionately involved in social justice issues, particularly those involving Polish independence, Hertha Ayrton was also a political activist. Ayrton became a supporter of women's causes. However, as she became older, Marie gave up her political activism for pure science. Hertha, on the other hand, supported the independence of Ireland and became more and more active in the English suffrage movement. Her fierce independence, her educational experiences, her husband's egalitarian attitudes, and her success at penetrating male-dominated institutions nurtured her inclinations. Her research, particularly that on the electric arc, while not theoretically important, was respected.

Even before Marie's troubles, she had planned to visit Hertha Ayrton during the summer of 1912. During the spring of 1911, Ayrton had visited Curie in Paris when Hertha presented her work on sand ripples to the société. After the presentation and a luncheon, Marie invited Hertha to visit her at her home in Sceaux and they discussed a future visit to England. Ayrton deplored those who had attacked Marie and sympathized with her in her health problems. She also congratulated her on her Nobel Prize. In a continuous spate of letters, Ayrton implored Marie to come to England with Irène and Eve. Ayrton explained that she would rent a house by the sea in Devonshire, so that Marie and her daughters would not have to go to London first. The trip was basically a success. Hertha was able to keep Marie's identity secret from the press and, although she was still often in pain, she managed to gradually increase her strength.

By the beginning of October, Marie was sufficiently strong to take the ferry from Dover to Calais in France and to continue on the train to Paris. Even though she was supposed to be careful, Curie quickly leapt into the scientific scene. She was upset with some of the work of Sir William Ramsay (1852–1916), who, like Curie, had published the atomic weight of radium. She complained to Rutherford that although they both arrived at the same results, he had the audacity to conclude that his work was the first valuable work on the subject. His comments about her experiments on atomic weights were unflattering and malicious, and she was furious.

There was also a problem regarding the radium standards. Rutherford was quite concerned because two individuals, a Viennese, Stefan Meyer, and Marie Curie, both had prepared primary radium standards. If each had worked accurately, the standards would be identical. If not, an unpleasant international incident could have occurred. While Marie was absent, Debierne set up the apparatus to test the standards against each other. To everyone's relief, the standards agreed.

During the first part of December 1912, Marie began her experimental work again. By this time the Langevin/Curie affair was over. To Irène's delight she dropped the name Sklodowska and again became Madame Curie. However, from 1911 to 1913 when Marie Curie was unable to concentrate on radioactivity, many new advances had been made.

Although Curie did not make any new discoveries herself during this period, she remained current on the many new concepts that were floating around. She and Einstein corresponded and she accepted many of the new ideas, including Rutherford's vision of the nuclear atom and Niels Bohr's quantum theory. Her own research during this time did not break any knowledge barriers and her involvement was mainly in the field of radiochemistry. Certainly her personal difficulties hindered her productivity. It must remain a matter of speculation whether she would have accomplished anything spectacularly new if her life had been different. Although Curie had many good friends in the scientific community, those whom she had offended were full of snide remarks and were contemptuous of her scientific abilities.

The year 1913 was much better for Curie than those immediately past. She attended the Solvay Conference in Brussels, traveled to Warsaw to dedicate a radium institute built in her honor, and went to Birmingham, England, to receive an honorary degree. She seemed to be less preoccupied with her health, although she was still too tired to spend as much time as she would have liked in the laboratory.

Marie began to entertain more often in her home. For example, she hosted Albert Einstein and his wife, Mileva, for a long visit in March 1913. In his thank-you note for the visit, Einstein was effusive about her hospitality. In fact, they so enjoyed each other's company that they planned a hiking vacation for the summer of 1913 in the Swiss Alps. The party included Curie and her daughters, Irène and Eve, their governess, and Einstein and his son, Hans. Marie was influential in assuring that Einstein got a job in Zurich, so he was somewhat in her debt. This did not stop him from being critical. Einstein claimed in a letter to his cousin, Elsa—whom he was courting to be his second wife—that the main way that both Marie and Irène expressed their feelings was by grumbling. Susan Quinn believes that Einstein was trying to assure Elsa that his outing with another woman was not any fun at all.

WORLD WAR

Many European countries were in the throes of setting up radium institutes. Their major interest was in the use of radiation for treating cancer. As the cost of radium skyrocketed, both physicists and medical researchers became more dependent on this precious commodity. The medical researchers were first in line to acquire radium, because of its potential to cure cancer. When Marie reported to the University of Paris in the 1912–1913 academic year, she complained tartly that the funds for basic research on radioactivity were disgracefully low. By this time, the Sorbonne seemed to have pushed the Langevin affair into the background and accepted the idea that the Pasteur Institute and the Sorbonne (University of Paris) should establish an institute devoted to the science of radioactivity. This institute was to be built on a new street honoring Pierre, the Rue Pierre Curie. As the workmen were apt to cut corners, Marie was vigilant in order to be certain that the building was built according to her specifications.

On July 31, 1914, the new institute was completed. However, it was not to be used as a laboratory for over four years, for World War I was about to begin. In August 1914, war mobilization began, followed by Germany's declaration of war on France. The men on the laboratory staff and the students were mobilized, leaving only Marie and the mechanic whose serious heart trouble kept him from joining the army. The French government was moved from Paris to Bordeaux, and many other Parisians followed—with many of the well-to-do going to the countryside.

Marie was forced to make a decision both about how to protect her family and about the gram of radium stored in the Rue Curie laboratory. About a week before the mobilization, Irène (sixteen years old), Eve (nine), and a Polish housekeeper and Polish governess went to the seacoast in Brittany for a holiday. Marie had planned to join them for a month's vacation. Their little fishing village of l'Arcouest was peaceful and they were surrounded by Marie's scientific friends, the Perrins and Borels. Although Irène was excited about the possibility of war, sensitive Eve was upset about the prospect. Realizing that she would not be able to get to l'Arcouest, Marie instructed Irène to do as the Perrins and Borels suggested. Irène tried to rebel and begged her mother to allow her to return to Paris. Realizing that they were better off where they were, on August 6, Marie wrote:

> My dear Irène, I too want to bring you back here, but it is impossible for the moment. Be patient. The Germans are crossing Belgium and fighting their way. Brave little Belgium did not allow them to pass without defending itself.[3]

As the danger of a German attack on Paris became more imminent, Curie had to protect her other child, the radium that was still in her laboratory. She was charged by the government to take it by train to Bordeaux (where the government in exile was hiding) for safekeeping. The radium was in a lead-protected bag that was so heavy that she could hardly lift it. After secreting the radium in Bordeaux, Curie returned to Paris on the train where people seemed gratified to see someone returning to Paris. When she arrived in Paris, she learned that the important battle of the Marne had begun. Terribly concerned about being separated from her daughters for so long, she, nevertheless, made the choice to remain in Paris at the institute. During the fierce fighting the French almost were defeated. However, the Paris taxis saved the day by rushing six thousand reserve troops to the front line. This battle was over on September 10, 1914. Although the French and British eventually won, from the standpoint of the loss of life, it was an unsatisfactory victory. The French had about a quarter of a million casualties, the Germans lost about the same number, and the British about thirteen thousand men. The importance of this battle was that the French and British forces were able to thwart the German plan for a speedy victory. Marie felt a great relief when, after that long battle, the French and British were victorious; however, because the German army retreated and was still capable of fighting, all prospects for a short war vanished. Nevertheless, this victory made it possible for Curie to bring Irène and Eve back from Brittany and continue with their schoolwork.

Marie was a loyal Frenchwoman by this time, but she never forgot that she was also a Pole. Poland was partially occupied by the Germans and she had not heard from her family. Marie desperately wanted to find something that she could do to aid the allied war effort. She chose a project that took her away from her beloved radium but one in which she was uniquely qualified to serve. Although she had never worked with Röntgen's x-rays, she had the theoretical knowledge to apply them to a practical use. Marie decided that her war work would consist of organizing radiology services for military hospitals.

By the beginning of World War I, physicians realized that x-rays would make visible the exact location of bullets, giving the wounded a greater chance to survive. Although at the beginning of the war the army health service had x-ray equipment in some of the large hospitals and even a few mobile units, it did not have units near the battle zones. The official view, shared by the front-line surgeons, was that there was no need for such facilities. The surgeons agreed with the official opinion because they had little confidence in the usefulness of radiology. Curie, on the other hand, was convinced that many lives would be saved. Through her passion, she was able to convince a private organization, the *Patronage des blesses,* to give her funds for the project. Using her official title, technical director of radiology, she located individual donors who saw the importance of the scheme. Even as the project developed, she still had difficulties with the army. It put roadblocks in her way at every turn, but on November 1, 1914, she finally received the needed permission.

Marie gathered together all of the apparatus that she could find in laboratories and in storage. Then she recruited and trained volunteer helpers to work in several stations throughout France. Although these stations were very useful, they were not sufficient to satisfy the need. Her solution was to outfit a radiologic car in collaboration with the Red Cross. Using an ordinary touring car she transported the radiologic apparatus. It included a dynamo worked by the engine of the car to furnish the electric current necessary to produce x-rays. This mobile unit could be available when any of the hospitals in the environs of Paris called.

Curie proved to be an efficient fund-raiser and established or greatly improved two hundred radiological installations. In addition, she was able to equip and give the army twenty radiologic cars, whose frames were donated by various people. Curie explained that these cars were especially important in the first two years of the war when the military possessed few radiologic instruments.

When mobilization first occurred, she wrote Irène that the two of them would try to make themselves useful. Irène beseeched her mother to allow her to return to Paris immediately. Marie replied that although she could use Irène's help it was still to dangerous for her to return. She admonished Irène to be patient and to look after her little sister. Looking after Eve was the last thing that Irène wanted to do. She continued to verbally assault her mother with ways in which she could help in the war effort. She suggested nursing with the Red Cross, serving as a secretary, or even teaching. Her peace of mind was not helped when her fellow teenagers accused her of being Polish and not a loyal French girl.

Continual nagging eventually wore her mother down, and seventeen-year-old Irène arrived in Paris at the beginning of October 1914 to help her mother. She had finished her preparatory studies and was ready to enter the Sorbonne but had to postpone her entry because of the war. Because she wanted to be useful, she studied nursing and learned radiology. Seeing the horribly injured young soldiers, many no older than Irène herself, must have been a shocking introduction to the horrors of war. Seeing the ambulance bring in the screaming blood- and mud-covered boys and men must have had a profound effect on Irène. However, she learned not only to be as detached as her mother but also how to deal with the officious military physicians who found it inconceivable that a woman could know more than they did.

Although Marie recalled that the conditions during these war years were especially difficult, she also noted that both she and Irène had pleasant memories of the hospital personnel, many of whom went out of their way to be helpful. Marie found that if she wanted the operation to go smoothly she had to look after each detail herself. She had to go through the bureaucracy to obtain passes and permission to move with her radiologic cars. She recalled:

> Many a time I loaded my apparatus on to the train myself, with the help of the employees, to make sure that it would go forward instead of remaining behind several days at the station. And on arrival I also went to extract them from the encumbered station.[4]

Curie's personal supervision ensured that the well-equipped cars were quickly assembled. The military chiefs were especially appreciative, because their appeals to the Central Health Service were answered, if at all, with snail-like speed. Finding competent radiologists to operate the equipment was more of a problem than obtaining the apparatus. Both the French and the English armies had recruited scientists without considering the potential loss of talented lives. Several of France's most talented scientists were killed. On the other hand, Curie recognized that wartime radiologic practice did not require a great deal of medical knowledge. She insisted that an intelligent person who had some idea about electrical machinery could be trained to be competent. She was especially pleased to train professors, engineers, and university students, insisting that they often made good manipulators. Another impediment was that she had to look for people who were not in the military or who were stationed close to the location where they were needed. As the war dragged on, the army recognized the need for more radiology technicians. It had even opened a school for x-ray technicians. Curie objected to the quality of the trainees, claiming that they were not selected because of aptitude and were often, at best, mediocre. The army finally agreed to ask Curie to conduct a course for technicians, but the facilities were so bad that she found another answer. Since women were not directly involved in fighting, her solution was to train women to do the radiologic work.

After the Health Service accepted Curie's proposal to add a radiology department to the newly founded nurses' school at the Edith Cavell Hospital, they began by training one hundred and fifty x-ray operators. Many of the young women had only an elementary education, but they were willing and able to undergo rather rigorous training. Not only did they have extensive practical training, but they also were given instruction in anatomy and provided with some theoretical principles. The teachers were volunteers, including Irène Curie. Even though the graduates were supposed to be aids to physicians, some showed that they were capable of doing independent work.

Curie's experience during the war led her to write a small book, *La radiologie et la guerre* (Radiology and the War). On the title page of this book, she calls herself "Mme. Pierre Curie," indicating that the long days of fearing that she had shamed the name of Curie were really past. In this book she stressed the importance of radiology, and compared its development during war to its peacetime uses. This book contains plates of radiographs. In the first chapter she described x-rays and the apparatus to be used in x-raying broken bones and finding foreign materials such as bullets embedded in the body; in chapter 2 she explained the procedure used to produce x-rays, and in the third chapter, discussed the x-ray installations in the hospitals. She then included a long chapter that outlined the radiological work in the hospitals, and in the last two chapters she discussed the radiological personnel. She ended the small book with an explanation of radiotherapy and radium therapy.[5]

Once Curie had her radiology stations set up and her radiological cars in action, she turned back to her true love, radium. In 1915, she retrieved the radium that she had deposited in Bordeaux and brought it back to Paris. This was the same radium that she and Pierre had originally separated. Since she had no time for pure research, she decided to use the radium for medical purposes. She had long been aware of the fact that radium was useful in treating cancer. She decided that it had special wartime potentials. It could be used to treat scar tissue, arthritis, and different ailments. Since radium itself had become immensely valuable, she did not want to risk the loss of any of the rare material. Thus the material that she placed at the disposal of the Health Service was not the radium itself but the emanation that it sporadically emitted. This emanation was the gas radon. If the radon was drawn off from the radium that formed it and was sealed in thin glass tubes, these tubes could be inserted into the body wherever they were needed, leaving the radium itself intact. Curie called these tubes emanation bulbs. Since she had no assistants she made most of the emanation bulbs herself. At this time she recognized that the emanations could have harmful effects, although she was certain that they were short term. Protecting her technicians was another reason that she prepared most of them herself. She insisted that the Health Board take special precautions to protect the laboratory where the bulbs were prepared from shells. As Curie noted, "the handling of radium is far from being free from danger (several times I have felt a discomfort which I consider a result of this cause)." Although she insisted that measures be taken to prevent the harmful effects of the rays on the persons preparing the emanation, she still thought that the discomfort, tiredness, and irritability that resulted from work with radium was temporary and would disappear as soon as the work ceased.

Amid the hard work of the war years, Marie obtained great pleasure in corresponding with the young soldiers. One of her favorite correspondents was her nephew, Maurice Curie, Jacques' son. Maurice wrote of the demoralized men who spent weeks, or sometimes even months, in frigid, soggy trenches infested with rats and lice. They were constantly exposed to enemy fire and had little opportunity to fight back. In April 1915, Germany introduced poison gas into the war. The French soldiers were forced to test gas masks in an enclosed room. Maurice wrote of the terrible headaches that resulted. Although Maurice survived the war, other young men with whom Marie corresponded were not as lucky. When the German army collapsed in the fall of 1918, the mood of the troops improved and a spirit of celebration was evident throughout France at the signing of the armistice on November 11, 1918. Marie herself celebrated with her friend Marthe Klein. After an unsuccessful search for flags, she bought red, white, and blue material and hurriedly produced homemade flags. She and Marthe joined the celebration by riding through the streets of Paris in her radiology car waving their flags.

After a war is over, as mothers, siblings, and wives mourn their dead, comfort their wounded, and repair broken relationships, people cannot help but ask the question "Why?" Why did we fight the war? Was it worth the price paid? After World War I, these questions were asked by many who were affected. Many people considered that both sides had much to answer for. Marie Curie, on the other hand, was convinced of the correctness of the Allied side. Perhaps her experience under the totalitarian Russian regime in Poland convinced her that living in freedom was the most important of values. She felt vindicated when her x-rays were able to save lives, confirming her views that pure science had the power to positively influence humankind.

The Treaty of Versailles, the treaty that ended the war, had in its essence the seeds of disaster. Curie, however, was gratified to find her native Poland again a sovereign state in its own right for the first time in 123 years. Other provisions of the treaty were less propitious. The formation of the League of Nations, meant to ensure that war was obsolete, failed to be ratified by the United States and without US support was a toothless tiger. The terms of the treaty forced Germany to reduce its armed forces drastically and did not allow it to use conscription. Many former German lands were given to Belgium, France, Denmark, and (of course) Poland. The land given to Poland became known as the "Polish Corridor" and it separated the main part of Germany from East Prussia. Germany lost all of its colonies. Most devastating for the future, Germany was forced to pay huge reparations to the Allies for the damages caused by the war and accept all of the blame for the war. New countries were created that upset the balance of power in Europe. Parts of two small countries were given to Italy and another new country was formed on the Adriatic coast called Yugoslavia, which included Serbia and Bosnia. Lithuania, Latvia, Estonia, and Finland were formed from the land lost by Germany's ally Russia. Czechoslovakia and Hungary were created out of the old Austro-Hungarian Empire. A new republican form of government, the Weimar Republic, was created based on proportional representation. Although it was intended to keep Germany from being taken over by a dictatorship, it led to the formation of more than thirty political parties. No one party was powerful enough to form a government on its own. Proud Germany was forced to its knees. These provisions led to utter humiliation. The people suffered under inflation, and when the opportunity arose to regain self-respect, Hitler appeared and promised them the moon and the stars; they were ready to follow him.

Even during the midst of the war, Marie made plans for the future. In 1915, the new laboratory building was completed, but she had no money or help to move the equipment into the new facility. Marie overcame that small obstacle by moving it herself with the help of Irène and the mechanic when he was not ill. One of her first concerns was to have trees planted on the laboratory grounds. She explained that she wanted to make things pleasant for those who would work in the new building. In addition to the trees, they planted beds of roses and other flowers. The planting and organization took several years, but all was completed by the beginning of the school year 1919–1920, when the country was demobilizing. During the spring of 1919, she planned special courses for American soldiers whom, she pronounced, "studied with much zeal the practical exercises directed by my daughter."[6]

The radiological services that were established during the war, such as the radiographic Nurses' School and the emanation service, were continued during peacetime as well. However, Paris was so devastated that money was not available for the laboratory work. Marie worried that she was no longer young. "I frequently ask myself whether, in spite of recent efforts of the government aided by some private donations, I shall ever succeed in building up for those who will come after me an Institute of Radium, such as I wish to the memory of Pierre Curie and to the highest interest of humanity."[7]

NOTES

1. *Nobel Lectures in Chemistry: 1901–1921* (Amsterdam: Elsevier, for the Nobel Foundation, 1966), p. 202.
2. Ibid., p. 203.
3. Eve Curie, *Madame Curie: A Biography* (Garden City, NY: Doubleday, Doran & Co., 1938), p. 290.
4. Marie Curie, "Autobiographical Notes," in *Pierre Curie,* trans. Charlotte and Vernon Kellogg (New York: Macmillan, 1923), p. 214.
5. Marie Curie, *La radiologie et la guerre* (Evreux: Imprimerie Ch. Herissey, 1921).
6. Marie Curie, "Autobiographical Notes," p. 221.
7. Ibid., pp. 223–24.

Chapter 10

MARIE AND THE UNITED STATES OF AMERICA

After the war, Marie Curie understood that most of her wishes—a stable France that would allow her to work on her science and a Poland without the heavy load of foreign occupation—were realized. The problems inherent in the peace settlement did not surface immediately after the end of the war in France. Armistice Day was celebrated by cheering throngs with the strains of the French National Anthem, "La Marseillaise," blasting through the streets. But for Marie Curie, it meant that she could get back to her scientific work. However, immediately after the war she took a much-needed vacation. After spending time at the warm coast, she returned to the laboratory, and science, refreshed. Her new laboratory was furnished sparsely. Even though she herself was perfectly happy with a Spartan laboratory, she wanted something very different for her workers. If the rest of the laboratory was to host a new French school of radioactivity, she would need sophisticated, expensive equipment. She would also need an additional supply of radium.

After pleading with many of the government agencies, she found her attempts to get funds fruitless. She found that one word—cancer—had the potential to bring in funds. Her international fame rested on her reputation as one who discovered a treatment for cancer. After World War I, an economically unscathed United States of America emerged as a world power. In 1920, women had won the right to vote in this country, and this precipitated events that led to an interview with an American women's magazine, the *Delineator*. The editor of this magazine, Marie Mattingly Meloney, known as Missy, had prodded writers who visited Paris to interview Marie Curie. Curie had her secretary turn every one of them away, explaining that she only discussed scientific matters. After Curie's unfortunate experience with the press over the Paul Langevin affair, she was very wary of journalists in any form. In desperation, the persistent Missy went to Paris herself. After agonizing over how to word a letter that would produce positive results and destroying ten unsatisfactory drafts, she finally wrote Marie the exact kind of note that produced results. Marie agreed to meet her for a brief interview.

Although she was a trained interviewer, Missy Meloney confessed that, when confronted with Marie Curie, she felt exceedingly timid. It was Marie who put Missy Meloney at her ease, rather than the other way around. During their discussion of radium, Curie explained that although the United States possessed about fifty grams of radium, France had only about one gram. Meloney quizzed her further and asked how much she herself had. Curie answered that she had none; the one gram belonged to her laboratory. Amazed, Meloney suggested royalties on her patents that would make her a rich woman. Missy felt shamed when Curie replied, "radium was not to enrich anyone. Radium is an element. It belongs to all people."[1]

In response to Meloney's question as to what she would choose if she could have anything she wanted, she replied that it would be a gram of radium. This question and its response led Meloney to research the price of a gram of radium. She found to her dismay that the market price was $100,000. She also found that although Curie's laboratory was almost new, it lacked sufficient equipment, and that the radium found there was only used for the extraction of the emanation (radon gas) of radium for hospital use in cancer treatment.

When radon gas was found in many homes in the United States, it received a considerable amount of bad publicity. Radon gas is produced by the natural disintegration of radioactive heavy metals such as uranium and thorium. As the atoms of radioactive heavy metals disintegrate, they change into increasingly lighter radioactive heavy metals until they end up as stable, nonradioactive lead. There are many ways in which naturally occurring radon can enter buildings. When radon gas is allowed to build up in an enclosed area such as a mineshaft or a basement, the radioactive hazard increases hugely because of the buildup of the products from the decay of radon gas. This product that we now see as dangerous was the same emanation that Marie Curie used for treating cancer.

Missy Meloney was impressed, as she knew she would be, by Marie. For her part, Marie was pleasantly surprised when she met Missy. The two women liked each other immediately, and a friendship was born. Through Marie, Missy saw the possibility of making a difference in the world; Marie saw Missy in the same way. Although they were similar in their goals and in some other respects, they also had important personality differences. Both had serious health problems and were slight of build. Missy, however, was an extrovert, albeit one who suffered bouts of depression, whereas Marie was a recluse. Ever since the Langevin fiasco, Marie had hesitated to confide in men. The degree of trust that Marie felt in Missy was so great that she confided the hurt that she felt from the Langevin affair. Missy was determined to raise the money to buy Marie her gram of radium. Convinced that there should be no difficulty in persuading wealthy American women to provide the bulk of the money, she resolved to convince ten wealthy women to give $10,000 each.

Marie Curie was still not popular in France even though she was recognized by some as the preeminent woman in that country. The Langevin incident still haunted her in her adopted home country. Meloney was convinced that if Curie would come to the United States, she could collect the gram of radium in person. Curie's distrust of the press convinced her that the American newspapers would leap upon the affair, and she would have to relive the previous horrible years. She confided her fears to Meloney, who assured Marie that she had nothing to fear from the American press. This statement almost backfired. Missy was able to put out the fires by visiting every leading New York newspaper editor and asking for their cooperation.

A series of articles in the *New York Times* indicate how well Missy had succeeded. This newspaper publicized the fund-raising activities, and as the time for the trip drew near the coverage increased. On February 7, 1921, a headline read "Radium Gift Awaits Mme. Curie Here." The subheadings continued "Prominent Women to Make Presentation to Foremost Woman Scientist on Visit to America" and "National Tour Is Planned." The announcement of the visit was made by Dr. Francis C. Wood of the Croker Memorial Cancer Research Laboratory. Although she was convinced of the practical use for radium, it was a fascination for pure science that motivated Marie Curie. However, her absorption with the basic structure of matter would not have appealed to the general public as much as radium's potential to fight the dreaded disease, cancer. This understanding was apparent in the way that the radium campaign was presented to the people and in their choice of a physician to announce the trip. The article also lists the twenty-one medical men on the committee, including Dr. Will J. May, president of the American Medical Association. At the time of this article, they still had not raised the money for the radium. Marie was described as

> fairly tall and slender, with a pale face. Her features are of the Polish type, the lips thin and suggesting the hardships which she and her husband endured in the difficult years before their great discovery. Her eyes are remarkable in their piercingness, her forehead of exceptional height, and her hair is golden and abundant.[2]

Two days later, Dr. Francis C. Wood corrected an impression from the February 7, 1921, article, which implied that the money for the radium had already been raised. He stressed that this "financial optimism" was "far from justified." He provided some optimism himself when he reported that there "is not the slightest doubt that the money will be obtained. But it has not been obtained as yet."[3] It seems obvious that he wanted to be certain that people would not stop giving, thinking that the money had already been raised.

Marie's approval of the gift plan was also reported in the *New York Times*. She wrote, "Permit me to thank you very sincerely for all the trouble taken with the object of securing for me a gift which would permit me to increase my work."[4] Reports with headlines such as "Cancer Deaths Here Are Increasing," in the March 7, 1921, *New York Times* helped spur interest in the fund drive. Even the president of the United States, Herbert Hoover, endorsed the movement to present a gram of radium to Curie. The New York Department of Health officials released a report citing a 6.6 percent increase in the number of cancer deaths in the city. Even more depressing to New York's citizens was the report that the rate of this city's increase was more than double that for the rest of the country. Cancer had surpassed tuberculosis as the chief cause of death from 1901–1921, having increased over 34 percent in that time period. With figures such as these, it was not surprising to find gifts to support what people regarded as a cure for this dread disease. The March 3, 1921, *New York Times* reported that a woman cured of cancer by radium contributed a $10,000 check for the Marie Curie fund. Other articles followed, all being very supportive of Curie and radium. One reported that the committee wanted the $100,000 to come from many small contributions rather than a few large ones. They made it clear that Curie never patented her processes and never exploited her products commercially. "She works for science, not for money, and it might be said with truth that she will be the trustee rather than the owner of this American gift."[5] As the time for Marie to come to the United States approached, the appeals became more insistent. Dr. Wood stated even more strongly than before that if the money was raised, Curie "proposes to devote her great scientific knowledge to the attempt to discover new methods for making radium more useful in the treatment of cancer. She is willing to devote her energy to the cause of suffering humanity. Will not the women of America make this possible?"[6] Still, by March 14, 1921, only $41,000 had been raised. As Marie was to sail on May 7, the situation was becoming more critical.

Missy Meloney was a complex person, combining social conservatism, hero worship, and an excellent business sense. In her publications, she chastised women who worked outside of the home and who left their children in the care of others. Missy herself had left her editorial post for ten years to care for her only child. In the same issue of the *Delineator*, which editorialized about radium for Marie Curie, there was an article by Vice President Calvin Coolidge railing against the radicals who infiltrated the women's colleges. And by radicals, he meant those who would break down the traditional roles of men and women. Meloney sent Curie a fictional book about children whose lives were ruined because their mother worked outside the home. Marie mildly protested the author's conclusion. Since Meloney worshiped Curie, she interpreted anything that Marie said that went against her own preconceptions as really agreeing with her. Missy commented on Marie's absence from her children during the war as a necessity that grieved her constantly. One of Marie's excuses not to come to the United States was that she would be away from her children. Missy considered this praiseworthy. However, if the truth were known, Marie's many absences from her children were choices that she made, not born of necessity.

Missy was a woman with a mission. The Marie Curie that she had constructed in her own mind and had dramatically presented to readers in America was not a real person. Nevertheless, Missy was so convinced that her idol did not have feet of clay that she managed to persuade others. At Missy's suggestion Marie wrote a brief autobiography. The myths perpetuated by Missy Meloney included the notion that Curie was herself still suffering from poverty. She was not. The most dangerous misconception proposed by Meloney was that Marie Curie would find a cure for cancer. Meloney was flamboyant and wallowed in overstatement. Nevertheless, she got what she wanted. Marie herself refused to be involved in the fund-raising, but wrote that if Missy was successful she would try to arrange to come to the United States to receive the gift. Missy overcame Curie's objections to traveling without her daughters by inviting Irène and Eve to come. The girls, of course, were delighted.

The news of how Marie was to be received in the United States led to the rehabilitation of her reputation in France. A fete was organized at the Paris Opera. She entered the auditorium amid thunderous applause and sat on the stage surrounded by some of France's most distinguished scientists. Sarah Bernhardt recited an "Ode to Madame Curie." The other entertainment was just as distinguished. France's press seemed to have forgotten the earlier scandal and members of the press core now almost tripped over one another to see who could supply the most elegant praise. After all the farewell ceremonies in France, Marie, Missy, and the two girls boarded the ship the *Olympic,* for their transatlantic crossing.

Missy and Marie had very different ideas about what the trip to the United States would involve. In Missy's eyes, the trip would involve whirlwind activities consisting of conferences, honorary degree ceremonies, and award acceptances. Marie, on the other hand, had envisioned a shorter and simpler visit. Although Missy thought she had scaled down the activities to satisfy Marie, they were only slightly curtailed. Missy had arranged for so many events that they could have been reduced in half and still have been too strenuous for Marie. The American public opened up its exuberant self to the drab little Polish/French woman. Plans began to take place long before the Curies actually sailed. They were to be met by the representatives of one hundred thousand college women. Representatives of the Association of College Alumnae and the Woman's University Club would host a reception on May 18, 1921, and each of the alumnae would be asked to donate $1.00 to the radium fund. Additional plans, without Marie's concurrence, had her staying in the United States for five weeks as a guest of the Marie Curie Radium Fund Committee. She was scheduled to visit cities of the East and the Midwest, as well as the Grand Canyon. The National Institute of Social Sciences was to award its gold medal to Curie on May 26, 1921. The award recognized her discovery of radium and its subsequent benefit to humanity.

On May 7, Marie, Irène, and Eve boarded the ship the *Olympic*. Their cabin was luxurious, but Marie, who had to be coerced by daughter Eve to buy several new dresses for the trip, preferred simpler surroundings. In a letter to Henriette Perrin, she explained that she felt some apprehension about leaving France "to go on this distant frolic, so little suited to my taste and habits." She found the crossing unpleasant, although she was not exactly seasick. By staying in her apartment she was able to avoid talking to curious strangers. Her daughters, however, were having a fine time. She also had high praise for Missy Meloney, who traveled with them and who "is as amiable and as kind as it is possible to be."[7]

The *New York Times* of May 11 announced their arrival. Not only had the Americans collected the $100,000, they had oversubscribed. If the fund continued to grow as it had been, the article suggested that a laboratory would be built for her on the outskirts of Paris under the direction of the University of Paris. When Marie finally arrived in New York, she was greeted by an intrusive swarm of paparazzi: journalists, photographers, and movie operators. Marie and her two daughters were dazed as countless mobs of curious people pushed and shoved each other in their attempt to see the woman whom the newspapers called the benefactress of the human race. One could distinguish Girl Scouts and schoolgirls as well as women who represented the Polish organizations of the United States. American, French, and Polish flags flew proudly.

The press hovered around Marie and her daughters. Her benefactors all wanted to entertain her. The day after the long voyage, she felt obliged to attend an event given in her honor. The curious press did not know the name of the hostess, but reported that Mrs. Andrew Carnegie's motorcar called for Marie shortly before noon and the assumption was that she was entertained at the Carnegie mansion. The next day she was even more tired and "denied herself to callers and remained at the home of Mrs. William B. Meloney." Meloney's house was filled with tributes from her admirers. A horticulturist whose cancer had been cured by radium sent Curie an enormous bouquet of roses that, he explained, "he had been cultivating and training for two months." Curie's hope for radium as a cancer cure had been misinterpreted by the press. She hastened to explain that radium was not a cure for all forms of cancer but was specific for certain forms.[8] As tired as she was, Marie and her daughters left Missy's house and set off for Northampton, Massachusetts, where she was to receive an honorary degree from Smith College.

Although most of the American colleges and universities were vying with each other for the privilege of awarding Curie honorary degrees, not all of the universities were enthusiastic. Charles Eliot, a former president of Harvard University, refused to meet Marie Curie in New York or to participate in a formal reception for her. Perhaps Harvard's Eliot was too influenced by Boltwood's assessment to offer her an honorary degree.

One US college, Vassar, was especially receptive to Madame Curie. She bestowed upon Vassar the singular honor of addressing its students and faculty on May 14, 1921. Edna Carter, chair of the Department of Physics at Vassar, explained that this was the only extended address that Curie made while visiting the United States. Carter prefaced the pamphlet that printed Curie's talk by reporting that "one realized how, closely environed by all the great realities of human experience, in the face of tremendous difficulties and with limited resources, she had pursued undaunted her search for truth."[9] For her part, Marie was very gracious in thanking the American women who made it possible for her to continue the research.

The strenuous American tour tired Marie. The buildup by Missy had been so enthusiastic that people expected a vibrant woman who would appreciate the boisterous American ceremonies. What they actually got was a tired, gaunt woman who avoided people and publicity whenever she could. Dowdy Irène was not popular with the public. She found the entire publicity circuit boring and used every opportunity to escape. Eve, on the other hand, loved the limelight. She was pretty, wore brightly colored clothing, and was pleasant to people. In spite of the troubles, Missy was able to not only raise enough money for a gram of radium but have over $50,000 to spare.

On May 19, the day before Curie was to be presented with the radium, Missy bestowed upon her the formal document. As Missy read the entire document aloud, Marie was displeased with what she heard. One sentence that was of vital importance to Marie had been left out of the document. This sentence involved the ownership of the radium and the rights of succession after her death. She insisted that the document be modified to read that the radium was "for free and untrammeled use by her [Curie] in experimentation and in pursuit of knowledge" and that it would become the property of her laboratory after her death.[10] Although it was late at night, she insisted that they must find a lawyer who would process the deed of gift. Among the witnesses was Vice President Calvin Coolidge's wife. At this point, it is possible that the American women who had worked so hard to help Curie felt a bit cheated. Marie was crotchety, stubborn, and insistent about getting her own way. Anticipating future trouble, they had the completed document translated into French. They still had not arrived at a solution as to the disposal of the leftover dollars. Marie insisted that since the money had been given in her name she, alone, should determine how it was to be used. Her benefactors thought that they should have some say in it, and the problem was not resolved until several years later. Marie, as usual, eventually triumphed.

The high point of the trip occurred on May 20, 1921, at 4:00 p.m., when President Warren Harding presented Mme. Curie with her gram of radium, really a facsimile. The actual radium remained in the factory. The presentation ceremony occurred in the East Room of the White House, where the French ambassador, Jules Jusserand, introduced Curie. In making the presentation, President Harding referred to

Curie as a noble creature, devoted wife, and loving mother who "aside from her crushing toil, had fulfilled all the duties of womanhood."[11] He praised her as "foremost among scientists in the age of science, as leader among women in the generation which sees woman come tardily into her own." Marie Curie responded to the president's presentation address "with a little speech of thanks delivered in broken English."[12]

If it had been up to her, Marie Curie would have cancelled the rest of the trip. She despised crowds and was feeling ill. She managed to escape a number of her obligations by having Missy cable the institutions that she was scheduled to visit, explaining that she was too ill to come. Irène and Eve attended the ceremonies and collected her honorary degrees in place of their mother. On May 23, Marie became so ill in Washington, DC, that she could not go to Philadelphia in time to be presented with two honorary degrees: one was an honorary doctor of law degree from the University of Pennsylvania, and the second was an honorary MD degree from the Women's Medical College. This MD degree was the first of its kind that Curie had received, although she already had accepted about sixty honorary degrees. Irène and Eve arrived in Philadelphia around noon, and Irène, clad in a black academic gown, accepted the degree from the Women's Medical College for her mother. In her simple acceptance speech she thanked the college for awarding her mother—"who was sorry not to be able to come here"—the degree.[13]

On May 25, 1921, the *New York Times* expressed concern about Marie's heavy schedule. Although those who planned the activities had only the best intentions in mind, they "subjected that distinguished woman to welcomes and honors so many and for her so heavy, that she is literally worn out." The article suggested that there should be a better way of welcoming and honoring a foreign visitor rather than subjecting the guest to endless receptions. It would have been much less strenuous for her if the academic degrees could have been conferred at a single ceremony. She then would have been free to follow her personal inclination "instead of being bound to a schedule of travel that covers the whole of every day for weeks on end."[14] Marie was ill off and on for most of the trip. Rumors were floating around that her illness was caused from handling radium, but she and the doctors concluded that it was only the strain of the trip. On May 29, it was announced that because of her exhaustion, all social plans for her upcoming trip to the West would be cancelled, and that the length of the trip would be shortened.

Missy herself was ill, but kept from Marie the fact that she was suffering from a possibly malignant tumor. In place of the previously scheduled hectic trip, she and Marie planned a quiet tour through the western United States, including the Grand Canyon. On June 2, they left New York for the western trip. It was apparent that the trip would not be as restful as proposed, when it was stated that she would visit a number of cities on her way back to the East. Her sought-after anonymity on the trip was impossible. When confronted by curious crowds, Marie came close to suffering panic attacks. The Curies and Missy gradually made their way back to New York. Curious to know her impression of the United States, the *New York Times* interviewed her the day before she was due to leave. The newspaper was gratified when she reported that "I feel that I have three countries—the land of my birth, the land of my adoption, and the land of my new friends." She praised the kindness of the people of the United States and said that she had only one regret about her trip—her physical inability "to do all the things I would wish to do and to meet all of the American people I much desire to meet."[15] The friendship between Marie and Missy had continued to strengthen, and they embraced with tears running down their cheeks as Marie and her daughters prepared to board the *Olympic* to return home.

On June 25, the *Olympic* departed with Marie Curie and her two daughters aboard. She told reporters that she felt refreshed by her visit to the United States. They arrived in Cherbourg, France, on July 2, where she was greeted by government officials and schoolchildren bearing flowers. The French seemed to have forgotten the scandal of years past. A squad of detectives took charge of the gram of radium and took it to Paris.

There were many benefits from Curie's trip to the United States. One of the most important ones was publicizing the idea that women could be scientists. The director of the Rockefeller Institute for Medical Research, Simon Flexner, appealed to young women who were graduating from Bryn Mawr to consider careers in the sciences. He noted that these careers were congenial, useful, and profitable. He referred to

Marie Curie as a woman who did important research. He did not, however, imply that all women were capable of original research. In fact, he stated that the majority of women had not developed a "scientific" mind. This lack was not, he asserted, caused by an innate inferiority to men but by a lack of education.

An American woman scientist, Christine Ladd-Franklin (1847–1930), who had studied at Johns Hopkins for four years and fulfilled the requirements for a doctorate there, was not awarded the degree from Hopkins until 1926, five years after Curie's visit. The publicizing of Curie's outstanding work may have made the Johns Hopkins faculty aware of the potential of women as scientists. On June 4, 1921, a letter from Ladd-Franklin appeared in the *New York Times* lauding the fact that Curie had been honored by having her name used as a common noun—curie, because of her distinguished contributions. The curie (the amount of radium emanation that is in equilibrium with one gram of radium) joined the watt, the ohm, the volt, the farad, the coulomb, the henry, and the ampere to honor the distinguished contributions of their discoverers. Marie Curie was the only woman to be so honored, but, Ladd-Franklin implied, not the last one.

When she returned to France, Marie noted that her work had been made easier by the gift of radium, and it inspired her to work all the harder to obtain more funding. She began to rethink her views about the relationship between the discoverer and the discovery. Both she and Pierre had been adamant in their views that science was for the benefit of humankind. An invention or a discovery did not belong to the scientist nor should he/she reap any material benefit from it. In her autobiography, Marie questioned the wisdom of that approach, but eventually concluded that it had been the correct one. When she observed others profiting monetarily from the fact that the Curies had openly shared their processes and techniques, she could not help but wonder if they had done the right thing. Industries were becoming rich because the Curies were morally opposed to taking out patents. Noting the high price of radium, she pondered the fact that in refusing to accept money, they had denied themselves and their children a fortune. And if they had guaranteed their rights, they could have had the financial means to support a satisfactory Institute of Radium. Yet even after this reconsideration, she was certain that they had made the correct ethical decision.

NOTES

1. Eve Curie, *Madame Curie: A Biography* (Garden City, NY: Doubleday, Doran & Co., 1938), p. 323.

2. "Radium Gift Awaits Mme. Curie Here," *New York Times,* February 7, 1921, p. 11, col. 3.

3. "Radium for Mme. Curie," *New York Times*, February 9, 1921, p. 8, col. 6.

4. "Mme. Curie Likes Gift Plan," *New York Times*, February 28, 1921, p. 15, col. 4.

5. "The Offer of a Gift Well Earned," *New York Times,* March 9, 1921, p. 12, col. 5.

6. "Appeals for Radium Fund," *New York Times,* March 10, 1921, p. 13, col. 4.

7. Eve Curie, *Madame Curie*, p. 326.

8. "Radium Not a Cure for Every Cancer," *New York Times,* May 13, 1921, p. 16, col. 3.

9. "The Discovery of Radium," address by Madame M. Curie at Vassar College, May 14, 1921.

10. Document dated May 19, 1921. *Biblioth que Nationale,* Paris, quoted in Robert Reid, *Marie Curie* (London: Collins, 1974), p. 263.

11. Ibid.

12. "Radium Presented to Madame Curie," *New York Times,* May 21, 1921, p. 15, col. 6.

13. "Honors for Mme. Curie Received by Daughter," *New York Times,* May 24, 1921, p. 19, col. 8.

14. "Worn Out by Her Welcome," *New York Times,* May 25, 1921, p. 16, col. 5.

15. "Mme. Curie Finds America a Marvel," *New York Times,* June 25, 1921, p. 11, col. 5.

Chapter 11

LAST YEARS

INSTITUT DU RADIUM

As was discussed in the previous chapter, the Radium Institute, begun in 1912, was completed during World War I. Every ounce of strength that Marie Curie could muster was put into making a flawless memorial to Pierre.

Before Pierre died, he had a vision of a laboratory dedicated to research in radioactivity. After a chair at the Sorbonne had been instituted in his honor, it seemed possible that this dream would materialize. Marie carried on his hope that such a laboratory would be established. The process of acquiring the laboratory was very convoluted. She had a strong advocate in the person of Émile Roux, the physician who was the director of the Pasteur Institute. This institute was well funded, and Roux conceived the idea that his institute could build the laboratory. The university agreed, and with Roux's help they planned two separate laboratories—one would be directed by Curie and funded by the university, and the other would be used to study the medical applications of radioactivity and funded by the Pasteur Institute. The second laboratory would be directed by the medical researcher Dr. Claudius Regaud. The two buildings would be constructed next door to each other and would together be known as the Institut du Radium.

Although Curie had never been interested in decorating or designing her own home, when it came to the laboratory she was a hands-on planner. Nothing was too insignificant. This laboratory was to be perfect. She met often with the architect and was not shy about letting her views be known. By the late 1920s the two original laboratories had expanded. Curie's laboratory had doubled in size, and the number of researchers had increased from only a few to thirty or forty. Regaud's facility had added an outpatient unit and another biology lab. Other expansion projects were planned.

After Curie returned from the United States to her own laboratory, her comfort level increased greatly. She was in her own surroundings. People she knew and loved supported her and she them. However, before she left on the trip she knew that her eyesight was failing and after she returned home it got considerably worse. Although she suspected that radium was the culprit that caused her cataracts, she refused to admit to her colleagues that there was anything wrong with her eyesight. Her daughters covered for her, as did select research persons whom she allowed to guide her through the streets of Paris. The people in her laboratory were well aware of the charade, but they permitted it to go on. Although she tried to cover her failing sight during her trip to the United States, the *New York Times* got wind of the problem. On July 1, 1921, the newspaper reported that an oculist, Dr. Peter A. Callan, had confirmed the report that she was threatened with blindness. Marie had visited his office a few days before she was to sail on the *Olympic*. Callan reported that a cataract on one eye was the problem, but that she was able to read with the other. He decided that immediate surgery was inadvisable because of her generally impaired physical condition. Presumably her health would have improved during the next six months, at which time the doctor assumed that she could undergo the ordeal of surgery. The cataract operation, which was much more serious in 1921 than at the present, would be performed in France. Her friends in the United States tried to protect her privacy by denying that her eyes were affected. They claimed that her visit to the oculist was only to have her glasses repaired.

Curie was very loyal to the students, technicians, and scientists who worked in her laboratory. She employed a number of women in her laboratory, including Ellen Gleditsch, Eva Ramstedt, Sybil Leslie, and Marguerite Perey. Many of the tasks delegated to the women were repetitive. Although they required precision and attention to detail, they did not demand a high degree of creativity. However, some of the women went far beyond the requirements of the jobs, including the four listed above. Although Marguerite Perey's (1909–1975) first responsibility in the lab was to wash test tubes, she advanced rapidly until eventually she discovered the element francium. Late in her life she became the first woman to be elected to the Académie des Sciences, an honor never enjoyed by Marie Curie. The Norwegian chemist Ellen Gleditsch (1879–1968), while working in Curie's laboratory, tested a claim by the British chemist William Ramsay (1852–1916) that radium was descended from uranium. She refuted his claim. She published numerous articles and several books on radioactivity. She also was the author of a textbook of inorganic chemistry and a biography of the chemist Antoine Laurent Lavoisier (1743–1974). May Sybil Leslie (ca. 1887–1937) was a British chemist who worked in Marie Curie's laboratory. While she was there she searched unsuccessfully for new radioactive elements in the mineral thorite. She then investigated thorium and its decay products. Eva Julia Augusta Ramstedt (1879–1974) was a Swedish physicist who worked in Curie's laboratory during 1910–1911. Among other accomplishments she found that the solubility of radium emanation (radon) varied with the solvent used and the temperature. She coauthored papers with Ramstedt and returned to Sweden where she worked at the Nobel Institute under Svante Arrhenius (1859–1927).

In addition to women, Curie's laboratory was ethnically and nationally diverse. There were people from many different countries. At least fourteen countries in addition to France were represented by researchers: Norway, Sweden, Russia, Poland, England, Germany, Belgium, China, Iran, India, Austria, Portugal, Switzerland, and Greece. All in all, the Curie laboratory was an exciting place to work. Marie Curie was the benevolent administrator, *la patronne,* who presided over this polyglot group.

It was becoming more and more difficult to ignore the harmful effects of radium. Even though it was not certain that Curie's cataracts were caused by her exposure, other effects were clear. The reluctance that kept Curie from admitting that her discovery could be harmful is understandable. She was unprepared to acknowledge that radium, her so-called child, could endanger people. As for the cataracts, respectable physicians were using radium to cure them. Today, physicians recognize that cataracts are an early sign of exposure to radiation.

As radium developed into an industry, something that Curie never anticipated nor desired, the evidence of deleterious effects began to accumulate. She recognized that certain local effects of radium could be harmful. She attributed the sores on her fingers to direct exposure to radioactive substances. But since the other symptoms experienced by those who came in contact with these materials varied greatly, she avoided naming radium as the culprit. Many of the workers in her laboratory complained about fatigue, but, she rationalized, this could be caused by any number of circumstances. She recalled that Pierre had endured excruciating pains in his legs, but never had felt the lethargy noticed by the other workers. Marie herself began to notice pains in her arms, but again, refused to blame radium. She did not advise the laboratory workers or the scientists who came to learn about radioactivity to take precautions.

Some of the reports emanating from London caused scientists and laboratory workers with radium to worry about their own health. In these reports radium was considered the cause of several deaths in a London hospital. After this initial report, other accounts began to snowball. One of the women who had worked with Curie in 1907, Norwegian scientist Ellen Gleditsch (1879–1968), was appointed to a committee to inquire about the effects of radium. Gleditsch asked Curie if France had established a similar committee to explore the effects of radium. Curie replied in the negative.

An incident that occurred in 1925 probably did more than any other to focus attention on the problem. A young woman from New Jersey, Margaret Carlough, worked in a radium dial-painting factory of the US Radium Corporation. It was common practice for the dial painters to point their brushes by using their lips. She sued the company, claiming that it had caused irrevocable damage to her health. On further exploration they found others suffering from what became known as radium necrosis (tissue death) and severe anemia (low number of red blood cells). This radium necrosis caused deterioration of the jaw. Approximately ten dial painters had died by the time Carlough sued the company, which denied that the deaths were caused by radiation. By 1928, fifteen deaths were attributed to radium exposure. Probably there were more misdiagnosed deaths. These deaths opposed all that was known about radium exposure. In the first place, the amount of radium in the luminous paint was minuscule—surely, they thought, not enough to harm a person. They also reluctantly found that when radium was ingested it was deposited in the bones of the arms and legs just as was calcium. From this vantage point it destroyed blood-forming tissues within the bones and had other harmful effects, including anemia and leukemia.

Missy Meloney wrote to Curie in June 1925 apprising her of the tragedy of the radium factory. The deaths of two engineers closer to home within four days of each other in Paris emphasized the problem. These two men worked for a factory that was preparing thorium X for medical purposes. Both had been ill for a considerable time. One, Marcel Demalander, was thirty-five years old and died of what was called severe anemia. The second man, Maurice Demenitroux, had been sick for a year and died of leukemia.

Each day brought additional reports of damage, including amputated fingers, blindness, and additional anemias and leukemias.

A report had been generated by the French Academy of Medicine in 1921 to address deaths reputedly caused by radiation. However, they concluded that most of the fears were unjustified, although they acknowledged that certain precautions should be taken when working with these materials. However, by 1925, a new report was generated by an investigative commission. Curie was on this commission, and by this time she was willing to recognize the potential danger in radioactive substances.

Radium as a cure for cancer had been on everyone's lips for several years. Unbeknownst to Curie, Missy Meloney had undergone radium treatment for cancer after Curie's return from the United States. Meloney explained that she had failed to inform Curie because the doctors had told her that it was an experiment. Another good friend, Loie Fuller, informed Marie that she had breast cancer. Most of the surgeons advised her to have a mastectomy, but one insisted that radium needles would give her an excellent chance for survival. She asked for Curie's opinion. All that Marie was able to do was to refer her to Regaud, who was the director of the medical part of the Radium Institute. Radium did indeed offer hope against cancer. The rays were able to kill cancer cells and keep them from dividing. However, cancer cells were not the only ones attacked by radiation. Other permanent damage on healthy cells could occur, although this was not known for several years.

The death toll from radium increased. Reports of deaths in radium laboratories and factories abounded. Although the evidence that radiation was the offender was becoming more and more compelling, neither Marie nor Irène was willing to admit that the evidence was unambiguous. After Irène, like her mother, received her doctorate from the Sorbonne, a reporter asked her about the dangers of radium. Although she admitted that she had already suffered from a radium burn, she insisted that it was not serious. Marie still was unable to admit to her own laboratory workers her fears about radium's effects. But at one point she acknowledged to a Polish laboratory worker, Alicja Dorabialska, that she did not fully understand radium's effect on health. She also confided to Dorabialska her fears that radium was the cause of her cataracts and the reason for her uncertain gait. As information accumulated about the health dangers from radiation, neither of her two daughters seemed overly concerned.

Irène had always been the favorite daughter. Her interests and personality were much like Marie's. Eve, on the other hand, found the arts and humanities more interesting than the sciences. She was an accomplished pianist and was interested in a career in journalism. She also was very softhearted, and when Marie Curie was confined to her bed after several unsuccessful cataract operations, Eve provided the support for her mother. She remained with her throughout the surgeries and stayed beside her bed reassuring her mother and convincing her that all was well. From this time on, Marie Curie had support from both of her daughters—Irène was her companion in scientific enterprises and Eve in domestic ones. One day, Irène dropped a figurative bombshell on her mother. She announced one morning at breakfast that she was engaged to be married and that the name of the intended groom was Frédéric Joliot.

The news of Irène's impending marriage came as a shock to Marie. Her elder daughter had been her mainstay in the laboratory, and the prospect of sharing Irène with someone else frightened her. However, when she found that the intended groom was Joliot, himself a brilliant scientist, she reluctantly accepted the marriage. Surely she recalled her early collaboration with Pierre Curie and recognized that her daughter was seeking the same kind of relationship. Irène and Frédéric later decided to hyphenate their names and became known as the Joliot-Curies. Some friends and colleagues feared that their disparate personalities would lead to an unhappy relationship. Marie Curie herself feared that the outgoing Frédéric would hurt the daughter in which she had invested so much of herself. Whereas Irène was forbidding and socially inept, Frédéric Joliot was a jolly man who loved to be around people. However, they had their science in common and it seemed to be enough. Frédéric was probably the more creative and brilliant scientist, but Irène was the better chemist and a skilled laboratory worker. Just as Marie and Pierre Curie had jointly won the Nobel Prize in 1903, Irène and Frédéric Joliot-Curie went on to win the coveted Nobel in 1935 for their work in artificial radioactivity. They were nominated for the physics Nobel Prize in 1934 but were passed over that year. Eve's husband, French diplomat Henry R. Labouisse, also won a prize, although not for scientific work. He accepted the Nobel Peace Prize for the United Nations Children's Emergency Fund in 1965.

As she grew older Marie Curie spent more and more time raising money for her laboratory and less in creative research herself. It was vitally important to her that her laboratory had sufficient funding to carry on Pierre's and her legacy. Missy Meloney was her indefatigable ally in the fund-raising efforts. Meloney traveled to Europe and all around the United States many times and solicited money whenever an opportunity arrived. The friendship between the two women deepened, and with Missy's aid, Marie herself developed a new appreciation of the importance of public relations to help with her money-raising goal. She visited many countries in Europe, giving lectures and indicating the necessity of money to continue research in radioactivity.

After Marie Curie had assured herself that the science of radioactivity in France was on secure footing, she turned her efforts toward Poland. Her sister Bronia had attempted to raise money for an Institute of Radium in Warsaw, but the results were disappointing compared to Missy's success for the French institute. Bronia's "Buy a brick for the Marie Sklodovska-Curie Institute" campaign had modest success but produced nothing like the amount needed. Marie Curie went to Poland, amazing the citizens of her native country with her still-perfect command of the Polish language. By 1925 the campaign had produced enough money so that a ceremonial cornerstone could be laid for the institute. However, even though the bricks metamorphosed into walls, and Marie and Bronia had contributed a good portion of their savings to the institute, it was still lacking the necessary radium for cancer treatments. Curie turned again to the generous Missy Meloney. Meloney managed to accumulate enough money for a second gram of radium. Meloney again mobilized a group of American women to raise money to buy a gram of radium. Missy Meloney's appeal to the American public for radium for Poland was less important than the support for Marie Curie herself. The personality cult of Curie was successful, and Poland got its radium. However, it is also true that the timing was providential. The money was raised before the stock market crash on October 24, 1929, which sent the entire country into a tailspin.

On October 16, 1929, Marie Curie arrived in the United States to thank its citizens once again for helping. Americans seemed to realize that it was modesty that caused her to shun the limelight. If she had had her own way, she would have entered the country "unheralded, as an ordinary traveler, and would seek to hide from the world the honors that lie in wait for her." Americans were not willing to let her hide, although they assured the press that she could go about her affairs without undue attention.[1] When she arrived on the French liner *Ile de France,* she sent word to a large contingent of reporters and photographers that her health would not allow her to be interviewed or photographed. The activities planned for her visit included the least possible number of tiring public appearances. Marie herself explained her limitations in a statement that she made upon her arrival.

> I am happy to come back to the United States. Deeply conscious of my debt to my friends in this country, it is with regret that I realize my physical limitations will make it impossible for me to do all of the things which my friends have been good enough to arrange for me. It will be a great pleasure for me to go to Washington again and to meet the President and Mrs. Hoover, and I especially want to attend the third annual dinner of the American Society for the Control of Cancer.[2]

In spite of the attempts to protect Marie from a hectic schedule, her popularity made it impossible to turn down certain invitations. One of these involved a visit to Schenectady, New York, where General Electric turned over its laboratories for her use for a day. She was invited to conduct any experiment that interested her. Nobody but laboratory assistants were allowed in the General Electric Laboratories during Mme. Curie's visit, and "she was absolute mistress of the plant."[3] Of course, one wonders what she could accomplish in a strange laboratory for one day, but it showed General Electric's respect for her and their attempt to "entertain" this bigger-than-life woman scientist. Students at St. Lawrence University also did their part to entertain Curie. She was serenaded by students singing as well as a thirty-piece student band. Although it was reported that she seemed to enjoy the serenade, on her way to still more activities, including another honorary degree, she suffered from fatigue.

This fatigue, as well as a drenching in the rain at the recent Edison celebration in Detroit, left her vulnerable to a cold when she returned to Missy Meloney's home in New York City. Although she was scheduled to attend the Roosevelt memorial dinner, she was forced to decline. During this visit she and Missy Meloney were guests of President Herbert Hoover and stayed in the White House for several days. They arrived in Washington on October 29, 1929, where they were met by the president's military aide at the Union Station who escorted them to the White House. In a letter to Eve she wrote that she had been given "a little ivory elephant, very sweet, and another tiny one. It seems that this animal is the symbol of the Republican party, and the White House is full of elephants of all dimensions."[4]

The presentation of the radium gift of $50,000 was made the next day, October 30, 1929. The gift itself was in the form of a bank draft encased in silver. Unlike eight years previously when President Harding presented a gram of radium to Mme. Curie, radium could no longer be bought in the United States. They would make the actual purchase in Belgium. In his presentation speech, President Hoover expressed appreciation for her service to humanity. He stated, "I am sure that I represent the whole American people when I express our gratification to Mme. Curie that she should have honored our country by coming here." Curie responded by noting that she was "conscious of my indebtedness to my friends in America, who for the second time, with great kindness and understanding, have gratified one of my dear wishes. My work is very much my life, and I have been made happy by your generous support of it."[5] She expressed her conviction that "scientific research has its great beauty and its reward in itself: and so I have found happiness in my work." The bonus, however was "an additional happiness to know that my work could be used for relief in human suffering."[6] She concluded by stating that she considered the American gifts of radium as a symbol of the enduring friendship that binds the United States to both France and Poland. After the official ceremony, she joined a magnificent reception and a dinner to celebrate—all of this went against her physician's insistence that she could only visit the United States if she would "attend no dinners, hold no receptions, and make no speeches." She broke her promise on all three counts![7] Of course, ever since she had arrived in the United States, she had attended dinners, receptions, and various forms of entertainment provided by her hosts. After celebrating her sixty-second birthday on November 7, Marie departed the next day on the French liner *Ile de France* for Europe. Before she left, she issued a personal statement expressing regret that her visit was over. She also apologized for not being able to accept all of the invitations from people and groups who wanted to meet her. She also stated that she would have liked to have visited more of the laboratories. On November 15, 1929, a tired Marie Curie arrived in France with little fanfare. She issued a statement of gratitude to the Americans for the gift of $50,000 and for the hospitality shown to her on her visit.

On May 29, 1932, Marie Curie visited Poland for the last time. At this time she inaugurated the Radium Institute of Warsaw. Bronia's organization and Marie's fund-raising for the radium had succeeded.

In spite of her increasingly fragile health, Marie refused to retire. Nevertheless she was far less active in the laboratory than she had been previously and the number of her new scientific papers diminished. Her scientific colleagues had long complained about her increased testiness and her decreased creativity as she aged. Her skills as a mentor, however, flourished, and she nurtured the young scientists who came to her laboratory. Newcomers in the laboratory were apprehensive when they first met this forbidding figure who appeared so cold and indifferent. Those who stayed, however, recognized that much of her apparent detachment was simply shyness and a lack of social skills. After several weeks in the laboratory, they realized that she saw them as being a part of a very special fraternity whose members were responsible for each other. They, after all, were the scientific elite. Many of these workers became intensely loyal to Curie. She was also extraordinarily helpful to the scientists whom she deemed worthy.

Gradually she was relinquishing some of her responsibilities as laboratory director to Irène and Frédéric, who were beginning a meteoric rise of their own. This occasioned some dissension among the laboratory workers as the Joliot-Curies were seen as heirs apparent to Marie's position. Marie Curie continued to give her course at the Sorbonne, and the students flocked to hear her just as they had during her first lecture immediately after Pierre's death. Although she recognized that her time as a creator of new theories had passed, she still played an important part in international scientific affairs. For example, one of the projects that she became passionate about was the question of scientific ownership and the rights of the scientists. When Pierre was alive she agreed with him that the scientist should not benefit from his or her discoveries. She did not benefit from the discovery of radium, insisting that scientific discoveries belonged to the world. Gradually she began to change these ideas, possibly because of her exposure to American science, where industry had often replaced individual achievement, and profit became a major motive for research. At any rate, while working with the League of Nations' International Committee on Intellectual Co-operation, she worked fervently on these ethical concerns. The same Marie Curie who declined any kind of patent rights from radium insisted in the 1930s that governments should reward creative scientists by establishing some type of royalty payment for those scientists whose discoveries benefited humankind. Marie Curie and Jean Perrin became activists on the part of scientists. Perrin insisted that without changes, France would remain a third-class scientific power, whereas Germany would be a first-class one. These ideas led to the formation of the French National Center for Scientific Research, the CNRS. This organization was important in determining the course of French Science in the future and Frédéric Joliot became an important guide in this organization.

As Marie became less able to work than she had previously, Irène and Frédéric Joliot-Curie took on additional responsibilities. Just as Marie and Pierre were partners, a new collaborative relationship was developed between her daughter and son-in-law. In the case of Marie and Pierre, both partners were withdrawn and diffident, whereas in the case of Frédéric and Irène two very different personality types were involved. Irène was like her mother, but Fred was outgoing and jolly. Nevertheless, both relationships were very fruitful. When interviewed in 1933, Frédéric explained that they compared notes and exchanged thoughts constantly. So constantly, "that we honestly don't know which of us is the first to have an original idea."[8] Irène nodded in agreement.

As Marie became more feeble, Irène and Frédéric Joliot-Curie had a daughter, Hélène, born in 1927; they later had a son, Pierre, born in 1932. Both children became physicists. Although Marie Curie adored her small granddaughter, she was as unemotional when dealing with the child as she had been with her own two daughters. However, as she grew older and more infirm, family became more important. Much of her enjoyment of her granddaughter was secondhand. She would watch Hela play at the beach but seldom entered into the fun. She still enjoyed swimming, and although she had spent many years of her life as a semi-invalid, she showed remarkable resilience.

An accident in the laboratory seemed to be the beginning of her rapid decline. In 1932, Marie fell and broke her right wrist. Although it was a simple fracture and should have healed quickly, it did not. After this accident her overall health began to deteriorate alarmingly. The trauma seemed to have caused the health problems, which had laid dormant, to surface. The radiation burns on her fingers became more inflamed. The thumping in her ears that she had experienced during her cataract problems emerged again. Her head hurt. For long periods of time she was confined to her bed. Marie Curie was convinced that she would not live much longer. She contacted her good friend Missy Meloney and invited, or rather demanded, that Missy visit her. The reason for her insistence was the recognition that her days were numbered. Always a private and somewhat diffident person, Marie confided to Missy Meloney her most fervent wish—that the radium would remain in her laboratory after her death and that Irène would inherit it. Practical Missy made the arrangements, and Marie was comforted in the realization that her precious radium would be preserved in the way that she wanted.

According to an article in the January 1, 1933, Sunday *New York*

Times Magazine, Marie Curie was still the active head of the Curie Institute. In Curie's eyes, this Institut du Radium was one of her greatest achievements, but one that was very expensive to maintain. Even in spite of her two lucrative American trips and funding from the French government, she was still dogged by money worries, for huge sums of money were necessary in order to keep the institute running. Consequently, she was constantly involved in fund-raising, a task that always made her uncomfortable. She had never really overcome her aversion to crowds and was always ill at ease on a lecture platform. The newspaper reported that "her many years of exposure to radium have had their effect on Madame Curie's health," but that she, nevertheless, continued to administer the institute, personally direct her staff's researchers, and gave two lectures a week as a professor of physics at the university. This was partially true in January 1933. Although she was often so fatigued that she could hardly drag herself to the laboratory, by pure will she did what was necessary as head of the institute.[9]

In December 1933, Marie suffered from a large stone in her gallbladder. Remembering that a gallstone operation had caused her father's death, she was determined to avoid surgery. Instead, she chose the alternative treatment of a strict diet and rest. She felt well enough to return to the laboratory, to design and build a country house at Sceaux, and to move to a new flat. Although she had eschewed luxuries during her entire life, as an aging, sick woman she did not hesitate to pledge herself to spend large sums of money for the luxuries that she had never before had.

Marie did not admit even to herself that her health was miserable. She went ice skating and skiing during the winter of 1934 with Irène and Frédéric and the children. Her actions, however, suggested that at some level she realized that death was looming. She explained to Irène where she could find the information that would serve as her will for the gram of radium. She also explained that the American documents were there, as well as a dossier, which included some letters. Marie, however, had destroyed all of the personal documents, especially those that were painful to her. These included all of the documents from Paul Langevin. She did keep Pierre Curie's old love letters to her as a young woman.

During Easter of the next year, Bronia visited Marie in Paris, and the two sisters went on an automobile trip to the south of France. Marie caught a bad chill and was both physically and psychologically in despair. Even so, her inner strength conquered, and when she returned to Paris she felt much better. But the improvement did not last. At times she was able to go to the laboratory. At others she felt dizzy and weak. In spite of her disabilities, she was cheered by the prospect of planning for her new villa. The bad days began to outnumber the good ones. She ran a fever almost constantly, and her body periodically would be racked by chills. Exhausted most of the time, she finally was unable to drag herself to the laboratory. The doctors diagnosed her problem as a recurrence of tuberculosis and suggested a stay at a sanatorium. Eve accompanied her mother and served as her nurse. Adhering to her mother's almost fanatical desire for privacy, Eve registered her at the sanatorium as Madame Pierre. By the time they arrived at the sanatorium, Madame Pierre's temperature had risen to 104 degrees, and she was so sick that she collapsed in Eve's arms. To make matters worse, they found that she did not have tuberculosis and blood tests showed that both her red and white blood cell counts were falling. The trip to the sanatorium that so sapped her strength was unnecessary. She was too weak to write and was forced to give up her long-standing habit of documenting all of the events in her life, although she was still able to read the lines on the thermometer.

To Eve and the doctors, it was evident that nothing could be done for her. Waiting in agony for her death, everyone was agreed that extreme methods to save her life would not happen. Marie herself did not contemplate the idea of dying. When on July 3, Marie Curie read the thermometer she was pleased to note that her fever had fallen. Although Eve assured her that this was the sign of her cure, it is actually the decrease in temperature that often precedes death. She began to hallucinate that her spoon was a delicate laboratory instrument. She would speak indistinctly of the things that had been a major part of her life and moved away from the humans whom she had loved and who had loved her. When Irène and Frédéric arrived she did not ask to see them. She fought hard not to let go of the life that was hanging by a thread. Her heart remained strong but the villain in the case, radium, proved stronger. On July 4, 1934, Marie Curie died at the Sancellemoz

Sanatorium. The attending physician, Dr. Tobe, wrote the following report: "The disease was an aplastic pernicious anaemia of rapid, feverish development. The bone marrow did not react, probably because it had been injured by a long accumulation of radiations."[10] The same culprit that had brought down her beloved Pierre so many years ago finally conquered the fragile appearing but strong woman who had given it life. In March 1956 Irène, too, was the victim of radiation, suffering from the same symptoms that killed her mother.

The American public had taken Marie to its heart, and the newspapers immediately reported the death of "their scientist." She had twice visited the United States. She had appreciated the obvious love and respect of the American people. Ironically, she had died on the fourth of July. A front-page article in the *New York Times* hailed her as a "martyr to science." Americans were surprised to hear of her death, because earlier newspaper articles had portrayed her as the active director of the Curie Institute. This article noted that "her death came as a surprise to all but her family and intimate friends, for the rare modesty of her character never deserted her and she did not allow the public to know how ill she was."[11]

At her death many scientists who had been critical of her work previously had nothing but kind words for her accomplishments. Albert Einstein, who was her friend and supporter although he was sometimes impatient with her personality, stated that she was "one of the most remarkable scientific personalities of our time." Praising her science he noted that "her ingenuity and her extraordinary energy enabled her to solve some of the most important problems which led to the discovery and to the scientific understanding of the radioactive phenomena." Although Marie and Einstein were always good friends, they sometimes disagreed over nonscientific matters. According to Einstein, she "was an unusually independent character," a statement that implied that her "independence" sometimes made it difficult for her to change her mind if she was certain that she was correct. Sometimes he and others did not come to the same solutions as she did regarding suitable ways to solve social and political problems. Her independence sometimes caused her to disagree with those in power as well as her friends, for she insisted on what she insisted was the best way to stand "wholeheartedly for justice and for progress in social matters."[12]

Physicist Dr. Robert Millikan (1868–1953), head of the California Institute of Technology, characterized Curie's discoveries of radium and the radiation emitted by it as "a starting point of the newer developments in physics." He contended that it was her ideas that convinced the world that "the heavens are not eternal and changeless, but that atomic transformations are taking place in nature all the time." He also paid tribute to her role in the cause for world peace as a member of one of the most important committees of the League of Nations.[13]

Sir J. J. Thomson, sometimes known as the "father of the electron," characterized Curie as one of the greatest physicists of modern times. After describing some of her accomplishments, the French scientist Duke Louis de Broglie (1892–1987) concluded by remarking "whether or not Mme. Curie died as a martyr to science, her loss is inestimable and will be mourned by scientists through the world."[14]

Many other scientists added to the plaudits, praising Marie Curie. Physicians represented another group who felt her loss very keenly. Many of those whom she had visited in the United States paid tribute to her work as "one of the foremost woman scientists and physicists of the world. Her passing out of the field of science has brought a great loss to the world."[15]

In accordance with the wishes of Marie and her family, the French government abandoned its plans for a national funeral. Marie Curie's body was brought to Paris on July 5, 1934, in the strictest privacy. On July 6, she was buried in a brief, simple ceremony without a civil or a religious service. She was buried in a plain oak coffin in the same grave where Pierre had been placed so many years ago. Roses were distributed to the gathered group, which consisted of 25 laboratory associates and 150 friends and scientists. Each mourner placed a rose on the closed coffin as he or she passed by. Not a word was spoken during the flower ceremony. Bronia and Józef came from Poland and threw some Polish soil into the open grave.

On April 20, 1995, the remains of both Pierre and Marie were transferred from the tomb in Sceaux to the Pantheon in Paris. Marie Curie was the first woman to be buried for her own accomplishments in France's national mausoleum.

NOTES

1. "Welcoming a Great Scientist," *New York Times,*" October 15, 1929, p. 30, col. 5.

2. "Mme. Curie Arrives. 'Happy to be Back,'" *New York Times*, October 16, 1929, p. 33, col. 3.

3. "Mme. Curie Examines Schenectady Plant," *New York Times*, October 24, 1929, p. 3, col. 5.

4. Eve Curie, *Madame Curie: A Biography* (Garden City, NY: Doubleday, Doran & Co., 1938), p. 343.

5. "Mme. Curie Receives $50,000 Radium Gift; Hoover Presents It," *New York Times*, October 31, 1929, p. 1, col. 3.

6. Eve Curie, *Madame Curie*, p. 343.

7. Ibid.

8. "Two Who Carry On the Curie Tradition," *New York Times Magazine*, January 1, 1933, p. 6, col. 4.

9. Ibid.

10. Eve Curie, *Madame Curie*, p. 384.

11. "Mme Curie Is Dead; Martyr to Science," *New York Times*, July 5, 1934, p. 1, col. 3.

12. "Scientists Mourn Mme. Curie's Death," *New York Times*, July 5, 1934, p. 16, col. 3.

13. Ibid.

14. Ibid.

15. Ibid.

CONCLUSION

To sum up Marie Curie's scientific achievements, we must address the relationship of her creativity to Pierre's. Did he supply the original ideas and she implement them? Was it significant that the original theoretical breakthroughs occurred during Pierre's lifetime? Her critics Ernest Rutherford and Bernard Boltwood sometimes had harsh words to say about Marie. Rutherford in a letter to Boltwood referring to her *Treatise on Radioactivity* (1910) wrote that

> in reading her book I could almost think I was reading my own with the extra work of the past few years thrown in to fill up. . . . Altogether I feel that the poor woman has laboured tremendously, and her volumes will be very useful for a year or two to save the researcher from hunting up his own literature; a saving which I think is not altogether advantageous.[1]

However, in an obituary notice in the British journal *Nature,* Rutherford was much more charitable. He noted that "she had long been regarded as the foremost woman investigator of our age."[2] Rutherford also praised her work as professor in the Sorbonne and as director of the Radium Institute in Paris and noted that she was actively engaged in research on the physical and chemical properties of radioactive bodies up until the time of her death.

Although Rutherford's assessment of Curie's achievements was not always flattering, Boltwood was even more negative, even vituperative, in his criticism of her work. When she received her second Nobel Prize, he was outraged because the theoretical work of Theodore Richards (1868–1928) on atomic weights had not been honored. According to Boltwood, Richards was much more deserving. Boltwood believed that Curie had received the award for stubborn perseverance rather than theoretical brilliance. In a letter to Rutherford he complained that "Mme. Curie is just what I have always thought she was, a plain darn fool, and you will find it out for certain before long."[3] The chemist George Jaffe, who visited the laboratory, assumed that it was Pierre "who introduced the ingenuity into scientific concepts . . . and the powerful temperament and persistence of Marie that maintained their momentum."[4] Mme. Curie was aware that critics proclaimed the originality in their work as her husband's.

In her 1911 Nobel speech, however, Mme. Curie made clear by her use of pronouns what she had contributed. The prize in chemistry was given to her "in recognition of her services to the advancement of chemistry by the discovery of the elements radium and polonium, by the isolation of radium and the study of the nature and compounds of this remarkable element."[5] She made it clear that she had the idea first when she said "some fifteen years ago the radiation of uranium was discovered by Henri Becquerel, and two years later the study of this phenomenon was extended to other substances, first by me, and then by Pierre Curie and myself."[6]

One of the most important theoretical assumptions surrounding radioactivity was the postulate that it was an atomic property. Although in her initial study she used the method of measurement invented by Jacques and Pierre Curie, it was the conclusion from the measurements that constituted the scientific originality. From the original publication it is not clear whether Marie or Pierre and Marie conceived the idea, for to them at that time it was obviously irrelevant. They concluded that the intensity of radiation is proportional to the quantity of material and that the radiation was not affected either by the chemical state of combination of the uranium or by physical factors such as light or temperature. This led to the important theoretical breakthrough that radiation was an atomic property. In Marie's 1911 Nobel Prize lecture she made it clear that this idea was hers. She explained that "the history of the discovery and the isolation of this substance has furnished proof of **my** [bolding mine] hypothesis that *radioactivity is an atomic property of matter and can provide a means of seeking new elements.*"[7] She also noted in the same lecture, "all the elements emitting such radiation **I** [bolding mine] have termed radioactive." This use of the first person was not used in her thesis (1902) where she described the creation of the hypothesis. In the thesis she merely wrote, "the radio-activity of thorium and uranium compounds appears an *atomic property.*"[8] However, she did note in her thesis that "**I** [bolding mine] have called radioactive those substances which generate emissions of this nature."[9]

If only Pierre's talks and writings were considered it would not be clear who had come up with the idea of radioactivity as an atomic property. When he presented the Nobel lecture of 1905 he did not designate individual roles. He said that radioactivity "presented itself as an atomic property of uranium and thorium, a substance being all the more radioactive as it was richer in uranium or thorium."[10] He wrote that "**we** [bolding mine] have called such substances *radioactive.*[11]

The hypothesis of the atomic nature of radioactivity motivated the long search that resulted in the isolation of polonium and radium. And the imaginative creation of a hypothesis distinguishes the outstanding scientist from the ordinary investigator. To be sure, Marie Curie's scientific genius had a second characteristic: perseverance. The labor necessary to substantiate her hypothesis was excruciatingly tedious and demanding. For Pierre it was unnecessary to go through the monotonous step-by-step chemical procedures to obtain pure radium, when, as a physicist, he could see what the results would be by applying his reason. Marie, on the other hand, also could hypothesize the results, but in order to persuade fellow chemists found it necessary to isolate the pure material no matter how long that it took. Her tenacity in the physical labor of attaining the pure material has contributed to the charge that her part in the Curie team was the less creative one. The evidence indicates, however, that in the discovery of radium Marie Curie contributed both the necessary hypothesis and the perseverance to demonstrate it in actuality.

In her later work the charge that Marie Curie was more involved in the minutiae of laboratory analyses than in creating new theories has more substance. Her insistence on isolating pure radium and pure polonium is a case in point. In her first effort to isolate radium, she had ended up with very pure radium chloride but not elemental radium. Lord Kelvin's suggestion in 1906 that radium was not an element but a compound of lead with a number of helium atoms had put her own work in jeopardy as well as that of Rutherford and Soddy, who theorized radioactive disintegration. Kelvin's ideas inspired Curie to embark upon a new series of purifications, being sure this time that the end product was the element radium. She also determined to settle the question of polonium as an element at the same time. Even though this process was eventually successful, and undoubtedly required skill and great patience, it did not involve additional theoretical suppositions, but it was admirable, nonetheless. Similarly the establishment of a radium standard in 1911, though an important achievement, was not predicated on additional theoretical assumptions.

Marie Curie's most scientifically creative years were indeed those during which she and Pierre shared ideas. Nonetheless, the basic hypotheses—those that guided the future course of investigation into the nature of radioactivity—were hers. Most of her later efforts were spent defending, elaborating, refining, and expanding these early ideas.

It is important to realize that her scientific creativity occurred in spite of all of the obstacles that she encountered, including prejudice because she was a woman, ill health for most of her life, loss of loved ones, and attacks on her personal life. Overcoming these barriers, she became not only the first woman to win a Nobel Prize, but the first person to win two Nobel Prizes. Add to that, the fact that her daughter also won a Nobel Prize. She was tenacious when she had a goal, and her life and work demonstrate that she is deserving of all of the accolades that she has received.

NOTES

1. Yale University Library. Ernest Rutherford to Bertram Boltwood, December 14, 1910, in Robert Reid, *Marie Curie* (London: Collins, 1974), p. 168.
2. Ernest Rutherford, "Mme. Curie," *Nature*, July 21, 1934, pp. 90–91.
3. Cambridge University Library. Bertram Boltwood to Ernest Rutherford, December 5, 1911, in Reid, *Marie Curie*, p. 213.
4. G. Jaffe, "Recollections of Three Great Laboratories," *Journal of Chemical Education* 29 (1952): 230–38.
5. *Nobel Lectures in Chemistry, 1901–1921* (Amsterdam: Elsevier, for the Nobel Foundation, 1966), p. 197.
6. Ibid.
7. Ibid., pp. 202–203.

8. Marie Curie, *Recherches sur les substances radioactives,* 2nd ed., rev. (Paris: Gauthier-Villars, 1904).
9. Ibid., p. 6.
10. *Nobel Lectures in Physics, 1901–1921*, pp. 73–74.
11. Ibid., p. 74.

EPILOGUE

In trying to present a subject as a complete person, a biographer is faced with a dilemma. Even though it is impossible to know the motives and thoughts of another person, the biographer is compelled to try to do so, with various degrees of success. Probably the most important thing for the writer to be aware of is his/her own biases, in other words, the tendency to emphasize or favor some aspects of a subject's life while giving less weight or possibly even excluding other aspects altogether. As careful and concerned as many biographers are to be objective, these biases often intrude. For example, in the biography of her mother, Eve Curie states that she had been completely objective in discussing Marie's personal life and scientific work, because she relied upon excerpts from her mother's actual words. However, the excerpts were clearly selected to show Marie in the most favorable light. Eve glossed over the fact that her mother often put her science before her responsibilities to her own children. Marie herself recognized the existence of this internal conflict when she wrote, "I have frequently been questioned, especially by women, of how I could reconcile family life with a scientific career. Well, it has not been easy."[1] Eve, understandably, wanted to present her famous mother in the best possible way, ignoring the darker side of her nature in favor of the more laudable aspects of this creative woman's life and character.

Perhaps not to such an extent as in Eve's biography, all biographers must grapple with their own individual biases to a greater or lesser degree when researching and writing about notable as well as notorious persons. The interpretation placed on an event or, since it is impossible and undesirable to embrace every experience in a subject's life, those events chosen for inclusion and discussion will reflect the author's particular interests if not his or her biases. For example, a biographer who is mainly interested in Marie's scientific contributions might de-emphasize the personal elements of her life while stressing her commitment to science. Or a biographer interested mainly in the more personal and private part of Marie's life might highlight this aspect while downplaying her approach to science. Others might concentrate on the political and social consequences of her life. As long as the biographer and reader understand the limitations of biography, it is an outstanding genre that can succeed in bringing a person to life while helping readers to understand his/her contributions.

One of the best ways to understand how Marie Curie considers her own life and how she wants others to remember her is to extract quotes from her own works. When comparing the relative importance of people and science, she wrote that "in science we must be interested in things, not in persons."[2] She also admonished researchers to "be less curious about people and more curious about ideas."[3] However, the biographer who looks beyond Marie's actual words to explore her life is rewarded with the insight that she was sensitive, easily hurt, and passionate about people as well as ideas. Marie Curie also elaborated on the potential and responsibility of science and scientists to improve the state of humanity.

> We cannot hope to build a better world without improving the individual.
>
> Toward this end, each of us must work for his own highest development, accepting at the same time his share of responsibility in the general life of humanity—our particular duty being to aid those to whom we can be most useful.[4]

Marie thoughtfully considered the nature of science itself. As scientists from the earliest of times have claimed, she also tells us that science is beautiful. "I am among those who think that science has great beauty. A scientist in his laboratory is not only a technician: he is also a child placed before natural phenomena which impress him like a fairy tale." She claimed that although searching for beauty and truth are the important goals, many useful applications may follow.

> Scientific work must not be considered from the point of view of the direct usefulness of it. It must be done for itself, for the beauty of science, and then there is always the chance that a scientific dicovery may, become like the radium, a benefit for humanity.[5]

Marie never wanted to profit from her research, although admitting that profit was motivating for some. Nevertheless she felt that society had an obligation to support the pure theoretical scientist whose work may not bear fruit for quite some time, if ever.

> Humanity certainly needs practical men, who get the most out of their work, and, without forgetting the general good, safeguard their own interests. But humanity also needs dreamers, for whom the disinterested development of an enterprise is so captivating that it becomes impossible for them to devote their care to their own material profit. Without doubt these dreamers do not deserve wealth, because they do not desire it. Even so, a well-organized society should assure to such workers the efficient means of accomplishing their task, in a life freed from material care and freely consecrated to research.[6]

In a letter to her sister Bronia in September 1927, Marie explained what science meant to her. She wrote that "sometimes my courage fails me and I think I ought to stop working, live in the country and devote myself to gardening. But I am held by a thousand bonds, and I don't know when I shall be able to arrange things otherwise. *Nor do I know whether, even by writing scientific books, I could live without the laboratory.*"[7]

In writing about Marie Curie we find a complex, not always consistent person. On one hand, she placed science on a pedestal declaring that ideas were more important than people, while on the other, she was nearly paralyzed with emotion after Pierre died. Throughout her life she was torn between rationality and sentiment when confronted with the illness and death of loved ones. When she herself became ill, she was in denial and tenaciously continued to work. While she disparaged those who profited financially from science, she felt that society had a responsibility to support scientists. As we consider Marie Curie's life, we can learn about creativity, love, despair, and hope all wrapped up in one extraordinary individual.

Education, especially scientific education, is the key to developing a prosperous society. However, since the 1960s scientific literacy in the United States has declined precipitously. Students in this country too often do not recognize science as an exciting and rewarding pursuit because they have had little exposure and less understanding of the field. "Difficult" and "boring" are descriptors too often used to portray this fascinating subject. One way to stimulate interest in science is to expose young people to the lives and achievements of scientists through biography. Reading scientific biography can help them realize how their world is a product of scientifically curious people from past eras.

For example, in the late nineteenth and early twentieth centuries society was transformed as scientific knowledge exploded. New elements and mysterious rays were discovered. Scientific theories were developed to explain the world in radically different ways. Nestled among the well-known, paradigm-changing male scientists such as Wilhelm Röntgen, Albert Einstein, and Niels Bohr we find Marie Curie, the iconic woman scientist who was the first woman to receive a Nobel Prize and one of only a few individuals to receive a second. Although she is best known for her part in the discovery of polonium and radium, her most important scientific idea was her theory that radioactivity was an atomic process. Curie's success is especially important to encourage girls that science is for them as well as for their brothers.

Although women have always had a role in the scientific enterprise beginning in the earliest of times, their achievements have been little recognized until recently. Through biographies of Marie Curie and other women scientists, we now know that women have played important roles in science. By including Marie Curie among the biographies of famous scientists girls and women may be inspired to embrace science as a possible profession.

NOTES

1. http://www.thinkexist.com/English/Author/x/Author_945_12/19/2010.
2. Eve Curie, *Madame Curie: A Biography* (Garden City, NY: Doubleday, Doran & Co., 1938), p. 350.
3. http://www.thinkexist.com/English/Author/x/Author_945_12/19/2010.
4. Eve Curie, *Madame Curie*, p. 53.
5. Ibid., p. 341.
6. Ibid., p. 78.
7. Ibid.

BIBLIOGRAPHY

PRIMARY SOURCES

Curie, Marie. *La radiologie et la guerre.* Paris: Librairie Félix Alcan, 1921.

———. *L'isotopie et les éléments isotopes.* Paris: Société de Physique. Paris: 1921.

———. *Oeuvres de Marie Sklodowska Curie.* Warsaw: Panstwowe Wydawnictwo Naukowe, 1954. This book contains most of Curie's papers in their original languages, French, Polish, German, and English. It also contains her doctoral thesis in French and Polish.

———. *Pierre Curie.* Translated by Charlotte and Vernon Kellogg. With an introduction by Mrs. William Brown Meloney and autobiographical notes by Marie Curie. New York: Macmillan, 1923.

———. "The Radio-Active Elements." *Independent. A Weekly Magazine* 15 (June 25, 1903): 1498–1503.

———. *Radio-active Substances.* London: Chemical News Office, 1903.

———. *Radioactive Substances.* New York: Philosophical Society, 1961. (English translation.)

———. *Radioactivité.* 2 vols. Paris: Hermann & Cie, 1935.

———. *Traité de radioactivité.* Paris: Gauthier-Villars, 1910.

SECONDARY SOURCES

Curie, Eve. *Madame Curie: A Biography.* Garden City, NY: Doubleday, Doran & Company, 1938.

Giroud, Françoise. *Marie Curie: A Life.* Translated by Lydia Davis. New York: Holmes & Meier, 1986.

McGrayne, Sharon Bertsch. *Nobel Prize Women: Their Lives, Struggles, and Momentous Discoveries.* 2nd ed. Washington, DC: Joseph Henry Press, 1998.

Nobel Lectures in Chemistry: 1901–1921. Amsterdam: Elsevier, for the Nobel

Foundation, 1967. Includes E. W. Dahlgren's "Presentation Speech," pp. 199–201, and Marie's Nobel Lecture of December 11, 1911, "Radium and the New Concepts in Chemistry."

Nobel Lectures in Physics: 1901–1921. Amsterdam: Elsevier, for the Nobel Foundation, 1967. Includes H. R. Törnebladh's "Presentation Speech," pp. 49–51; a biographies of Pierre, pp. 79–80 and Marie, pp. 82–83; and Pierre's Nobel lecture of June 6, 1905, "Radioactive Substances, Especially Radium," pp. 73–78.

Ogilvie, Marilyn. "Curie, Marie (Maria Sklodowska)." In *The Biographical Dictionary of Women in Science*. Vol. 1. Edited by Marilyn Ogilvie and Joy Harvey. New York: Routledge, 2000, pp. 311–17.

Opfell, Olga S. *The Lady Laureates: Women Who Have Won the Nobel Prize*. Metuchen, NJ: Scarecrow Press, 1986, p. 203.

Pycior, Helena M. "Pierre Curie and 'His Eminent Collaborator Mme. Curie' Complementary Partners." In *Creative Couples in the Sciences*. Edited by Helena M. Pycior, Nancy G. Slack, and Penina Abir-Am. New Brunswick, NJ: Rutgers University Press, 1996.

Quinn, Susan. *Marie Curie: A Life*. New York: Simon and Schuster, 1995.

Reid, Robert. *Marie Curie*. London: Collins, 1974.

Weill, Adrienne R. "Curie, Marie (Maria Sklodowska)." *Dictionary of Scientific Biography*. Vol. 3. Edited by C. C. Gillispie. New York: Charles Scribner, 1971, pp. 497–503.

INDEX